Lecture Notes in Mathematics

A collection of informal reports and seminars
Edited by A. Dold, Heidelberg and B. Eckmann, Zürich

Series: Institut de Mathématique, Université de Strasbourg
Adviser: M. Karoubi and P. A. Meyer

284

Paul-André Meyer

Université de Strasbourg, Strasbourg/France

Martingales and Stochastic Integrals I

Springer-Verlag
Berlin · Heidelberg · New York 1972

AMS Subject Classifications (1970): primary: 60 G 45
secondary: 60 H 05

ISBN 3-540-05983-0 Springer-Verlag Berlin · Heidelberg · New York
ISBN 0-387-05983-0 Springer-Verlag New York · Heidelberg · Berlin

Offsetdruck: Julius Beltz, Hemsbach/Bergstr.

Introduction

This volume contains a preliminary version (the second draft) of the first two chapters of a book on martingales and stochastic integrals. If the following chapters are ever to be written, it will be as a joint work with Cl. Dellacherie : we plan to write at last the chapters on continuous time martingales in 1972-73, and to publish them in the Lecture Notes series. The next installment then will be on stochastic integrals. Formal publication will probably take place much later, and at that time the first chapters will probably (and hopefully) be obsolete.

There exist several texts on martingale theory, standard or specialized ; at least three more are forthcoming, which should be rather outstanding. On the other hand, nothing exists in book form on the general theory of stochastic integrals, and our emphasis will be on that side. The book would just have been called that way, hadn't it been that the title already belongs to the wonderful little book by McKean (which however deals only with the sample continuous case). Accordingly, I have mentioned only the most standard facts about convergence, but I tried to be rather complete about transforms, quadratic variation, etc. The point of view here is quite different from that of my own chapters [1] on martingale theory, which tried to follow the ideas of probabilistic potential theory.

Part of the material was taught (in Gerglish) in 1970/71 at the Albert-Ludwig University at Freiburg i/Breisgau. During that year I was subsidized by the Deutsche Forschungsgemeinschaft, and during the following Wintersemester by the University itself. I am most grateful to them for this generous and pleasant hospitality. This stay in Germany, and the idea of writing the book for a German publisher, explain why I didn't write in French . Some English speaking friends complained about it. Let them face the facts : there is a worldwide speculation on the devaluation of English.

For corrections and improvements to these chapters I owe many thanks to P.Assouad, S.D.Chatterji, K.L.Chung, Catherine Doléans-Dade (and Everett), C.Dellacherie, J.L.Doob, N.Kazamaki and J. Walsh. For allowing me to make use of unpublished material, to A. Garsia , R.Getoor and M.Sharpe , and to J.Neveu , who allowed me to borrow two beautiful proofs (II, n°s 47 and 53) from his forthcoming book on martingales.

My warmest thanks, finally, go to Prof. F. Eicker, who invited me to Germany, and helped me friendly and patiently in my difficulties with everyday life and the German language.

P.A. Meyer

Contents

---

* All definitions, theorems, remarks ... are sequentially numbered in each chapter. References throughout the book, and in this table, concern these numbers (not page numbers).

FIRST PART : MARTINGALE THEORY

This part consists of three chapters : fundamentals, the discrete case, and the continuous case. We have tried not to duplicate the similar section on martingale theory in [1]. We have done our best to modernize the contents, taking into account a number of new proofs which have appeared since, and the new emphasis on martingale transforms. Also, we have added the new inequalities of BURKHOLDER on the quadratic variation, and some other modern material - though so much work is in progress that our choice can hardly be considered definitive.

CHAPTER I . BASIC DEFINITIONS

§ 1 . Some elementary facts

1. We assume that our reader is familiar with the basic definitions of probability theory. Let $(\Omega,\underline{\underline{F}},P)$ be a complete probability space, and let $\underset{\sim}{T}$ denote either $\underset{\sim}{N}$, the set of all positive[1] integers, or $\mathbb{R}_+=[0,\infty[$. We say that a family $(F_t)_{t\in\underset{\sim}{T}}$ of sub-σ-fields of $\underline{\underline{F}}$ satisfies the usual conditions if $\underline{\underline{F}}_t$ increases with t, contains all sets of measure 0, and if in the continuous case (i.e., when $\underset{\sim}{T}=\mathbb{R}_+$) the family is right continuous

$$\underline{\underline{F}}_t=\underline{\underline{F}}_{t+}=\bigcap_{s>t}\underline{\underline{F}}_s$$

a property which has no counterpart in the discrete case.[2] We shall usually denote by $\underline{\underline{F}}_\infty$ the σ-field $\bigvee_{t\in\underset{\sim}{T}}\underline{\underline{F}}_t$ generated by $\bigcup_{t\in\underset{\sim}{T}}\underline{\underline{F}}_t$.

2. A process is a family $(X_t)_{t\in\underset{\sim}{T}}$ of real valued random variables on Ω (the values ±∞ are excluded, unless specifically stated). We generally use capital letters to denote processes, and the same letter with the time t in index to denote their random variables. Thus if a sentence begins with " Let X be a process...", the reader should keep in mind that X_t will denote the value of X at time t.

3. A process X can also be considered as a mapping $(t,\omega)\mapsto X_t(\omega)$ from the product set $\underset{\sim}{T}\times\Omega$ to $\mathbb{R}$. If this mapping is measurable with respect to the product σ-field $\underline{\underline{B}}(\mathbb{R}_+)\times\underline{\underline{F}}$ (in the continuous case), the process will be called measurable. In the discrete case, every process is measurable.

---

[1] Positive in this book always means ≥0 .

[2] Unless explicitly stated, we always assume the usual conditions below, in the continuous case.

4. Probabilists usually do not distinguish between random variables which are a.e. equal. The corresponding notion for processes is the following : probabilists cannot distinguish two processes X and X' such that, for a.e. ω, the sample functions $X_{\bullet}(\omega)$ and $X'_{\bullet}(\omega)$ are identical. Quite naturally, X and X' are said to be <u>indistinguishable</u> . In the discrete case, this means only that $X_n = X'_n$ a.s. for each n, but in the continuous case the property that $X_t = X'_t$ a.s. for each t is much weaker than indistinguishability, since there are uncountably many t. A process which cannot be distinguished from 0 is often said to be <u>evanescent</u> . The same word may be applied to a subset of $\underset{\approx}{T}\times\Omega$ whose indicator is an evanescent process.

5. Let us consider a typical example of the discrete case : a gambler plays at times 0,1,...,n,... and we want to study the evolution of his bank account. To describe the situation, we shall in most cases need two processes (at least) : the first one, which we denote by V, describes the amount of money bet on the first... n-th... gamble. The second one X, describes the outcomes of the successive gambles (its values are not necessarily numbers, they may be horses or cards). It is clear that V and X haven't the same status, since the outcome of the gamble is (generally) not known by the time the gambler bets : V_n must be known " just before" time n, and X_n " at " time n. We are thus led to the following definitions.

<u>Discrete case</u> . A process X is said to be <u>adapted</u> to the family $(\underline{\underline{F}}_n)$ if X_n is $\underline{\underline{F}}_n$-measurable for every $n \geq 0$. A process V is said to be <u>predictable</u> (w.r. to the family $(\underline{\underline{F}}_n)$) if V_0 is $\underline{\underline{F}}_0$-measurable, and for every $n>0$ V_n is $\underline{\underline{F}}_{n-1}$-measurable.

<u>Continuous case</u>. We shall say as above that a process X is <u>adapted</u> if X_t is $\underline{\underline{F}}_t$-measurable for every $t \geq 0$, but this definition isn't precise enough. Let us denote by $\mathcal{L}$ (resp. $\mathcal{R}$) the set of all processes U, adapted to $(\underline{\underline{F}}_t)$ and such that for a.e. ω the sample function $U_{\bullet}(\omega)$ is left continuous on $]0,\infty[$ (resp. right continuous on $[0,\infty[$). Let us denote also by $\mathcal{P}$ (resp. $\mathcal{W}$) the smallest σ-field on $\mathbb{R}_+\times\Omega$ with respect to which the elements of $\mathcal{L}$ (resp. $\mathcal{R}$) are measurable. Then we say that a process V is <u>predictable</u> if the

3

mapping $(t,\omega)\longmapsto V_t(\omega)$ is $\mathcal{P}$-measurable, and that X is *well-measurable* if $(t,\omega)\longmapsto X_t(\omega)$ is $\mathcal{W}$-measurable (the word *optional* may also be used). The elements of $\mathcal{L}$ (resp. $\mathcal{R}$) will be simply called adapted left continuous (right continuous) processes.

In the discrete case, we shall occasionnally use the word well-measurable as synonymous to adapted, in order to reach a complete analogy with the continuous case. Though the discrete case generally is " elementary" , while the continuous case uses " advanced" methods (capacity theory), the analogy between the two cases is very fruitful and illuminating.

MARTINGALES AND SUPERMARTINGALES

6. A process X is a *supermartingale* (w.r. to the family $(\underline{\underline{F}}_t)$) if it is adapted, if $E[|X_t|]<\infty$ for every t , and if X satisfies the *supermartingale inequality*

(6.1) for $s<t$, $X_s \geq E[X_t|\underline{\underline{F}}_s]$ a.s.

or, more explicitly, if for every $A\in\underline{\underline{F}}_s$

(6.2) $\int_A X_s\, P \geq \int_A X_t\, P$.

We say that X is a *submartingale* if -X is a supermartingale, a *martingale* if X and -X are supermartingales, i.e. if equality holds in (6.1) and (6.2).

The definitions thus are exactly the same in the discrete and the continuous case - but in the latter case the class of interesting supermartingales will be restricted to the right continuous ones. The above definitions can be generalized in many directions, but we shall not attempt it at this stage. We just state two trivial results.

7.THEOREM. *Let* X *and* Y *be two supermartingales relative to the same family* $(\underline{\underline{F}}_t)$. *Then* $aX+bY$ ($a\geq 0$, $b\geq 0$) *and* $X\wedge Y$ *are supermartingales*.

The second result rests on the property that every convex function c on $\mathbb{R}$ is the upper envelope of a countable family of linear (affine) functions .

8.THEOREM. *Let* X *be a martingale, and* c *be a convex function on* $\mathbb{R}$. *If the process* $c\circ X$ *is such that* $E[|c\circ X_t|]<\infty$ *for all* t, *then* $c\circ X$ *is a submartingale* (*same result if* X *is a submartingale and* c *is convex increasing*).

The most important cases are $c(t)=|t|^p$ for $p\geq 1$, and $c(t) = |t| \log^+|t|$.

STOPPING

9. A stopping time (or optional r.v.) of the family ($\underline{\underline{F}}_t$) is a random variable S on Ω, taking values in $\underset{\sim}{T}\cup\{+\infty\}$, such that

(9.1) for every t, the event $\{S\leq t\}$ belongs to $\underline{\underline{F}}_t$.

In the discrete case, this amounts to $\{S=n\}\in\underline{\underline{F}}_n$ for all n. On the other hand, in the continuous case it is sufficient to check that $\{S<t\}\in\underline{\underline{F}}_t$, since we assume that ($\underline{\underline{F}}_t$)"satisfies the usual conditions", and hence is right continuous. Similarily, the fact that $\underline{\underline{F}}_0$ contains all null sets implies that any r.v. which is a.s. equal to a stopping time is a stopping time.

If S and T are stopping times, so are $S\wedge T$ and $S\vee T$.

10. A σ-field $\underline{\underline{F}}_S$ can be associated to any stopping time S in the following way : $\underline{\underline{F}}_S$ consists of all $A\in\underline{\underline{F}}_\infty$[1] such that, for every t,

(10.1) $\qquad A\cap\{S\leq t\} \in \underline{\underline{F}}_t$

The events in $\underline{\underline{F}}_S$, intuitively, are those which are" known at time S" or rather (because of the assumed right continuity) " just after time S " We shall define later (n° I.34) the σ-field $\underline{\underline{F}}_{S-}$ consisting of all events which are known " just before time S".

We may replace in (10.1) $\{S\leq t\}$ by $\{S<t\}$, because of right continuity, but (10.1) is the convenient definition for the general case of a family which isn't necessarily right continuous.

The following properties are trivial, and left to the reader (or look at the proofs in [1], IV, n^{os}36-42) S and T denote stopping times .

a) S is $\underline{\underline{F}}_S$-measurable.

b) $S\leq T \Rightarrow \underline{\underline{F}}_S\subset\underline{\underline{F}}_T$.

c) $A\in\underline{\underline{F}}_S \Rightarrow A\cap\{S\leq T\} \in \underline{\underline{F}}_T$.

---

[1] This restriction that $A\in\underline{\underline{F}}_\infty$ isn't always made, and isn't very important. We have decided in n° I.1 that $\underline{\underline{F}}_\infty$ denotes $\bigvee_t \underline{\underline{F}}_t$, but we might have denoted instead this σ-field by $\underline{\underline{F}}_{\infty-}$, and decided that $\underline{\underline{F}}_\infty$ ought to be any larger σ-field, thus loosening the restriction.

11. Let H be a random variable, taking values in $\underset{\sim}{T}$ (and hence finite valued in the cases $T=\underset{\sim}{N}$ or $\mathbb{R}_+$). Let X a process. We denote by X_H the mapping $\omega \mapsto X_{H(\omega)}(\omega)$ on Ω. In the discrete case, it is very easy to check that X_H is a random variable. In the continuous case, this is true <u>if</u> X <u>is a measurable process</u>, since X_H is the composition of the measurable mappings $\omega \mapsto (H(\omega),\omega)$ into $\mathbb{R}_+\times\Omega$, and $(t,\omega) \mapsto X_t(\omega)$. If X is right continuous, let us denote by D_n the set of all dyadic numbers $k2^{-n}$ ($k\in\underset{\sim}{N}$), and by $H_n(\omega)$ the smallest element of $D_n \cap]H(\omega),\infty[$; then H_n is a countably valued random variable, and decreases to H as $n\to\infty$. The right continuity (a.s.) of X implies that $X_{H_n} \longrightarrow X_H$ a.s., a way of " computing" X_H which is sometimes useful.

12. For instance, assume that H is a stopping time. Then every H_n is a stopping time, and $\underline{\underline{F}}_H = \bigcap_n \underline{\underline{F}}_{H_n}$ (easy proofs : see [1], IV, n°s 42-43). If X is adapted and right continuous, an elementary computation shows that X_{H_n} is $\underline{\underline{F}}_{H_n}$-mesurable, and therefore X_H is $\underline{\underline{F}}_H$-measurable. Using a monotone class argument, this result is easily extended to all well-measurable processes (see also [1], IV, n°s 47-50).

13. We are now ready to show the connection of stopping times with stopping ! Consider a stopping time H and a process X. Then the <u>process</u> X <u>stopped at time</u> H , often denoted by X^H, is defined as follows

(13.1) $$X^H_t = X_{t\wedge H}$$

If X is adapted and right continuous, then X is right continuous, and adapted to the families $(\underline{\underline{F}}_t)$ and $(\underline{\underline{F}}_{t\wedge H})$. A monotone class argument then shows that if X is well-measurable w.r. to $(\underline{\underline{F}}_t)$, then X^H is well-measurable w.r. to both families $(\underline{\underline{F}}_t)$ and $(\underline{\underline{F}}_{t\wedge H})$.

14. The most important example of stopping times is that of <u>first hitting times</u>. Let X a well-measurable process, and let I be a Borel subset of the real line. Define the first time X hits I (this isn't a generally accepted terminology !) as follows

(14.1) $$D_I(\omega) = \inf \{ t : X_t(\omega)\in I \} \quad (\text{as usual, } \inf \emptyset = +\infty)$$

It is quite easy to see in the discrete case that D_I is a random variable and a stopping time.In the continuous case the same result is true, but its proof requires the following theorem :

Let $(\Omega,\underline{\underline{G}},P)$ a complete probability space, and let A be a subset of $\mathbb{R}_+\times\Omega$, belonging to the product σ-field $\underline{\underline{B}}(\mathbb{R}_+)\times\Omega$. Then the projection of A on Ω belongs to $\underline{\underline{G}}$.

This is an easy consequence of the capacity theorem of CHOQUET. See for instance [1], III, n°24[1]- the proof that the projection is $\underline{\underline{G}}$-analytic is given (somewhat obscurely !) in the proof of IV.52. This result has a curious history . Assuming the fact (which was well known to the Polish mathematicians) that A can be constructed from $\underline{\underline{G}}$ with the help of SUSLIN'S operation (A), it may be said that it was known long before CHOQUET : see SAKS [2], chapter II § 5 , the theorem was proved in 1918 ! It is not advisable now to look for proofs without capacities, since they would yield only the above result, and not the deeper ones we shall need later.

Given this result, let us prove that for every r $\{D_I<r\}$ belongs to $\underline{\underline{F}}_r$ (i.e., D_I is a stopping time). This set is the projection on Ω of $\bigcup_n A_{r-1/n}$, where

$$A_s = \{ (t,\omega) : X_{t\wedge s}(\omega) \in I \}$$

The mapping $(t,\omega) \mapsto X_{t\wedge s}(\omega)$ is $\underline{\underline{B}}(\mathbb{R}_+)\times\underline{\underline{F}}_s$-measurable : this is very easy if X is adapted and right continuous (see [1],IV.47), and extends to any well-measurable X through the usual monotone class argument. Thus A_s belongs to $\underline{\underline{B}}(\mathbb{R}_+)\times\underline{\underline{F}}_s$, $\{D_I<r\}$ is the projection on Ω of an element of $\underline{\underline{B}}(\mathbb{R}_+)\times\underline{\underline{F}}_r$, and we can apply the above result from measure theory with $\underline{\underline{G}}=\underline{\underline{F}}_r$.

TRANSFORMS (DISCRETE CASE)

15. Let us come back to the simple example of n°5. The successive bets of the gambler at the 0-th, 1-st,...,n-th.. game are $V_0,V_1,\dots,V_n,\dots$ On the other hand, consider a gambler which plays exactly the same games as the first one, but each time betting an amount of money equal to 1. Let y_n be his return from the n-th game, so that his total return just after the n-th gamble is $Y_n=y_0+y_1+\dots+y_n$. Then the total return of the first gambler after the n th game is

$$Z_n=V_0y_0+V_1y_1+\dots+V_ny_n = V_0Y_0+V_1(Y_1-Y_0)+\dots+V_n(Y_n-Y_{n-1}) .$$

The interesting process is Z, and this representation is a nice way to disentangle, in many cases, the contribution to Z of the size of the bets, and that corresponding to the outcomes of the gambles.

[1] Or better [5], chapter I, theorem 32.

For instance, at a coin tossing game, the y_i are the usual BERNOULLI random variables, and it is a quite natural and interesting problem to study the processes Z which arise in that way from suitable choices of V.

However, we aren't specially interested in gambling, and we shall just give for further use the following definition :

DEFINITION . Let V and Y be two processes, the first one predictable and the second one adapted (w.r. to the family $(\underline{\underline{F}}_n)$). Then the process

$$Z_n = V_0 Y_0 + V_1(Y_1 - Y_0) + \dots + V_n(Y_n - Y_{n-1}) \tag{15.1}$$

is called the transform of Y by V, and denoted by $Z = V \cdot Y$.

The fundamental importance of this simple notion for martingale theory was discovered by BURKHOLDER (see [3]). Our main purpose in this book consists in extending to the continuous case, under the name of stochastic integrals, the definition of transforms.

If S is a stopping time, Y an adapted process, the stopped process Y^S (n° 13) is just the transform of Y by the predictable process $V_n = I_{\{S \geq n\}}$.

DOOB'S DECOMPOSITION (DISCRETE CASE)

16. THEOREM. Let $X = (X_n)_{n \in \mathbb{N}}$ be an adapted process such that $E[|X_n|] < \infty$ for all n. Then X can be decomposed as

$$X = Y - A \tag{16.1}$$

Where Y is a martingale, A a predictable process such that $A_0 = 0$. This decomposition is unique (i.e., unique up to evanescent processes). It is called the DOOB decomposition of X (see [4],p.296)

PROOF. First let us show the uniqueness. Consider two such decompositions $X = Y - A = Y' - A'$. Then $A - A' = Y - Y' = M$ is a predictable martingale with $M_0 = 0$. The martingale property means that $E[M_{n+1} | \underline{\underline{F}}_n] = M_n$ a.s., the predictability that $E[M_{n+1} | \underline{\underline{F}}_n] = M_{n+1}$ a.s., hence $M_{n+1} = M_n \dots = M_0 = 0$ a.s., and M is evanescent.

As for existence, we set

$$\begin{array}{ll} A_0 = 0 & Y_0 = X_0 \\ A_1 = A_0 + E[X_0 - X_1 | \underline{\underline{F}}_0] & Y_1 = Y_0 + (X_1 - E[X_1 | \underline{\underline{F}}_0]) \\ \dots & \dots \\ A_n = A_{n-1} + E[X_{n-1} - X_n | \underline{\underline{F}}_{n-1}] & Y_n = Y_{n-1} + (X_n - E[X_n | \underline{\underline{F}}_{n-1}]) \end{array}$$

A and Y obviously have the desired properties.

17. If X is a supermartingale, we have $E[X_{n-1}-X_n|\underline{\underline{F}}_{n-1}] \geq 0$ for all n, and therefore $A_0=0$, $A_n \geq A_{n-1}$ a.s. (we shall say in this case that A is an <u>increasing process</u>[1]). Conversely, if X=Y-B, where Y is a martingale and B is increasing and adapted (not necessarily predictable) , with $E[B_n]<\infty$ for all n, then X is a supermartingale. Y and B aren't unique unless predictability of B is assumed.

The extension of DOOB's decomposition to continuous time was an open problem for a while. We shall deal with it in chapter III.

18. It will be useful to have names for the processes appearing in the decomposition.

Let X be any process (not necessarily adapted) such that $E[|X_n|] <\infty$ for all n. Define A as in n°16 . The process -A then will be called the <u>compensating</u> (or <u>balancing</u>) process for X, and denoted by $\tilde{X}$ or $X^{\sim}$. The process $X-\tilde{X}$ = X+A is denoted by X^c, and called the <u>compensated</u> (or <u>balanced</u>) process of X. Note that X^c isn't equal to the process Y of n°16 unless X is adapted.

The case of an adapted X is the most important one. Then X^c is a martingale, and $\tilde{X}$ is the only predictable process such that $\tilde{X}_0=0$ and $X-\tilde{X}$ is a martingale. The non adapted case occurs from time to time, specially with non adapted increasing processes.

Finally, let X and X' be two processes with $E[|X_n|]<\infty$, $E[|X'_n|] <\infty$ for all n. We say there are <u>associated</u> , and write $X \sim X'$, if they have the same compensating process : for instance, $X \sim \tilde{X}$. In the adapted case, this means that X-X' is a martingale.

19. One last word about notation : we shall not set a rigid rule, but we shall try to use capital letters for processes, and <u>the same lower case letter</u> for the corresponding difference process :

$x_0=X_0$, $x_n=X_n-X_{n-1}$; $X_n=x_0+x_1+\ldots+x_n$.

This is logically inconsistent : how could we denote the difference process of x ? However, as a practical rule it may help the memory.

With these notations, we have if X and X' are as in n°18

(19.1) $\tilde{x}_0 = 0$, $\tilde{x}_n=E[x_n|\underline{\underline{F}}_{n-1}]$ for n>0

(19.1) $X \sim X' \iff E[x_n|\underline{\underline{F}}_{n-1}] = E[x'_n|\underline{\underline{F}}_{n-1}]$ for n>0 .

20. Let A and B be two associated processes, a and b be the corresponding difference processes, V a predictable process such that V_na_n and V_nb_n are integrable for every n (if A and B are increasing one may just assume that V is positive). We have for n>0

[1] " Increasing" is meant to imply "adapted" unless explicitly stated.

$E[V_n a_n | \underline{\underline{F}}_{n-1}] = V_n E[a_n | \underline{\underline{F}}_{n-1}] = V_n E[b_n | \underline{\underline{F}}_{n-1}] = E[V_n b_n | \underline{\underline{F}}_{n-1}]$.

Otherwise stated, <u>V.A and V.B are associated</u>. In particular, if A and B are increasing with $A_0=B_0=0$, associated, and T is any stopping time, then $E[A_{T\wedge n}]=E[B_{T\wedge n}]$ for all n. Letting $n\to\infty$ we get that $E[A_T]=E[B_T]$, with the natural definition $A_\infty = \lim_n A_n$, $B_\infty = \lim_n B_n$.

This trivial result will allow an easy proof of a useful theorem, namely LEVY's sharpening of the BOREL-CANTELLI lemma :

21. THEOREM. <u>Let</u> (b_n) <u>be a sequence of positive integrable random variables</u> (<u>not necessarily adapted</u>). <u>Set</u>

$a_0=E[b_0|\underline{\underline{F}}_0]$, $a_n=E[b_n|\underline{\underline{F}}_{n-1}]$, $B_n=b_0+\dots+b_n$, $A_n= a_0+\dots+a_n$.

<u>Then</u> $\{A_\infty < \infty\} \subset \{B_\infty < \infty\}$ <u>a.s.</u>. <u>Conversely , if the sequence</u> (b_n) <u>is adapted and</u> $b^*= \sup_n b_n$ <u>belongs to</u> L^1, <u>we have</u> $\{B_\infty < \infty\} \subset \{A_\infty < \infty\}$ <u>a.s.</u> . <u>Note that if</u> $b_0=0$ <u>we have</u> $A=\widetilde{B}$.

PROOF. A and B are associated, and we can reduce to the case $b_0=a_0=0$. Since A is predictable, T= inf { n : $A_{n+1}>N$ } is a stopping time, and A_T is dominated by N. Hence $E[B_T]=E[A_T]<\infty$ (n° 20) and B_T is a.s. finite. Thus B_∞ is a.s. finite on $\{T=\infty\}=\{A_\infty \leq N\}$, and letting $N\to\infty$ yields the first result.

To prove the converse assume that B is adapted, and define T = inf { n : $B_n>N$ } . Then B_T is dominated by $N+b^*\in L^1$, and therefore integrable. Hence $E[A_T]$ is finite, and we reason as above.

22. Why is this called a BOREL-CANTELLI lemma ? Take for (b_n) a sequence of indicators of independent events β_n, for $n>0$, define $b_0=0$, $\underline{\underline{F}}_0=\{\emptyset,\Omega\}$, $\underline{\underline{F}}_n$ for $n>0$ being the σ-field generated by $b_0,\dots,b_{n-1}$. Then a_n is just $P(\beta_n)$, and the set $\{A_\infty<\infty\}$ is Ω if $\sum P(\beta_n) <\infty$, $\emptyset$ if it is equal to $+\infty$. Then theorem 21 means that

if $\sum P(\beta_n)$ converges, then $\sum b_n<\infty$ a.s. (the trivial part !)

if $\sum P(\beta_n)$ diverges, then $\sum b_n= +\infty$ a.s. .

This is exactly the classical BOREL-CANTELLI lemma .

§ 2. Predictability in continuous time

The preceding section contained only those definitions and results on continuous time processes which were trivial extensions of the discrete situation (there was just one noteworthy exception in n°14). Here we are going to see a good deal of the so called "general theory of processes", a theory concerned with the behaviour of processes under suitable measurability assumptions, and whose main tools are capacity theorems.

The reader is advised to read this section after chapter II (since its results are used nowhere in chapter II), and before chapter III.

Let us recall that the family $(\underline{\underline{F}}_t)_{t\in\mathbb{R}_+}$ satisfies the usual conditions (n°1), that processes are considered as mappings on $\mathbb{R}_+\times\Omega$ (n°3), that $\mathcal{L}$ and $\mathcal{R}$, the adapted a.s. left continuous (right continuous) processes generate respectively the predictable σ-field $\mathcal{P}$ (the well-measurable σ-field $\mathcal{W}$) on $\mathbb{R}_+\times\Omega$ (n°5). Our first task will consist in finding more explicit ways of describing these σ-fields.

GENERATORS FOR THE FIELDS $\mathcal{P}$ AND $\mathcal{W}$

23. DEFINITION. Let S and T be stopping times such that $S\leqq T$. We denote by [S,T] the set

$$\{ (s,\omega)\in\mathbb{R}_+\times\Omega : S(\omega)\leqq s\leqq T(\omega) \}$$

If S=T, we write [T] instead of [T,T]. The open stochastic intervals]S,T[and the half open ones]S,T],[S,T[have similar definitions.

We are dealing with subsets of $\mathbb{R}_+\times\Omega$, not $\overline{\mathbb{R}}_+\times\Omega$: thus it isn't quite correct to call [T] (as we shall do) the graph of T . Another small inconsistency is the fact that, if S and T happen to be constants s and t, [s,t] denotes both an interval of $\mathbb{R}$ and a subset of $\mathbb{R}_+\times\Omega$. That isn't really too bad [1].

24. THEOREM. Every well-measurable set belongs, up to an evanescent set, to the σ-field generated by all intervals [S,T[.

Every predictable set belongs (up to an evanescent set) to the

[1] DELLACHERIE adopts in [5] the notation [[S,T]],[[S,T[[... for stochastic intervals.

σ-field generated by all sets $\{0\}\times H$ ($H\in\underline{\underline{F}}_0$) and $]u,v]\times H$ (u,v numbers with $u<v$; $H\in\underline{\underline{F}}_u$).

PROOF. Note first that the above sets $]u,v]\times H$ are stochastic intervals $]S,T]$: set $S=u$ on H, $+\infty$ on H^c , $T=v$ on H, $+\infty$ on H^c.

The indicator functions of stochastic intervals $[S,T[$ belong to $\mathcal{R}$, those of sets $\{0\}\times H$ or $]u,v]\times H$ as above , to $\mathcal{L}$. We are going to prove that, conversely, every element of $\mathcal{R}$ or $\mathcal{L}$ can be approximated by linear combinations of such indicator functions, at least on the complement of some evanescent set. This will prove the theorem. We begin with the simpler case of a process $X\in\mathcal{L}$: since we may neglect evanescent sets, it isn't a restriction to assume that all sample functions $X_\cdot(\omega)$ are left continuous. Set first

$$X_0^n=X_0 \quad , \quad X_t^n= X_{k/2^n} \text{ for } \frac{k}{2^n} < t \leq \frac{k+1}{2^n} \qquad (k\in\underline{N})$$

It follows from left continuity that the process X^n converges to X as $n\to\infty$. Let Y^n be the n-th LEBESGUE approximation of X^n

$$Y^n(t,\omega)= p/2^n \text{ if and only if } \frac{p}{2^n}<X^n(t,\omega)\leq \frac{p+1}{2^n} \quad (p\in\underline{Z})$$

Then $|X^n-Y^n|\leq 2^{-n}$, hence Y^n converges to X as $n\to\infty$. It is a trivial matter now to check that Y^n is a linear combination of indicators of the desired kind.

The proof for well-measurable processes is a little more difficult. Let X belong to $\mathcal{R}$; as above we shall suppose (neglecting an evanescent set) that all sample functions $X_\cdot(\omega)$ are right continuous. Taking $\varepsilon=2^{-n}$ we define the following functions by transfinite induction

$S_0=0$

$S_1(\omega) = \inf \{ t : |X_t(\omega)-X_0(\omega)|>\varepsilon \}$

.....

If α is any ordinal[1] number, we define

$S_{\alpha+1}(\omega) = \inf \{ t>S_\alpha(\omega) : |X_t(\omega)-X_{S_\alpha}(\omega)| > \varepsilon\}$

($+\infty$ if this set is empty, in particular if $S_\alpha(\omega)=+\infty$). If α is a limit ordinal, i.e. isn't of the form $\beta+1$, then we set

$S_\alpha(\omega) = \sup_{\beta<\alpha} S_\beta(\omega)$.

The fact that S_α is a stopping time for every α is now proved by transfinite induction, using n°11 (which gives the measurability

[1]Here, as everywhere in this book, we use loosely the word " ordinal" to mean " countable ordinal"

12

of X_{S_α} w.r.t. to $\underline{\underline{F}}_{S_\alpha}$, on $\{S_\alpha<\infty\}$) and n° 14[1] (to get that $S_{\alpha+1}$ is a stopping time – the case of a limit ordinal is trivial, since the supremum of a countable family of stopping times is a stopping time). Next we use the following obvious fact about transfinite numbers : if f is a monotone mapping from the set of ordinals to $\mathbb{R}$, then there exists at least a (countable !) ordinal γ with $f(\gamma)=f(\gamma+1)$, since the set of open disjoint intervals $]f(\alpha),f(\alpha+1)[$ is countable, and the set of all ordinals uncountable. We apply this to the function $f(\alpha) = E[\exp(-S_\alpha)]$: the ordinal γ is such that $S_\gamma = S_{\gamma+1}$ a.s.. On the other hand, right continuity implies that $S_\gamma<S_{\gamma+1}$ on $\{S_\gamma<\infty\}$, and from these two properties together we deduce that $\{S_\gamma=+\infty\}$ has probability 1.

Remembering that ε was taken equal to 2^{-n}, we denote by X^n_t the process $\sum_{\alpha<\gamma} X_{S_\alpha}(\omega)I_{[S_\alpha,S_{\alpha+1}[}(t,\omega)$. The definition of S_α implies $|X^n_t(\omega)-X_t(\omega)|\leq 2\varepsilon$ for $S_\alpha(\omega)\leq t<S_{\alpha+1}(\omega)$, hence $|X_.(\omega)-X^n_.(\omega)|\leq 2\varepsilon$ on $\{S_\gamma=+\infty\}$. As in the first part of the proof, denote by Y^n the n-th LEBESGUE approximation of X^n_t : if we neglect an evanescent set, X and Y^n differ by 4.2^{-n} at most. It remains to show that Y^n is a countable linear combination of indicators of stochastic inter $[S,T[$, and this amounts to the following : if S,T are two stopping times with $S\leq T$, if $A\in\underline{\underline{F}}_S$, then the set

$$\{(t,\omega) : \omega\in A,\ S(\omega)\leq t<T(\omega)\}$$

also is a stochastic interval $[U,V[$: this is trivial. One just sets $U=S_A$, $V=T_A$, using the following very important notation.

25. DEFINITION. <u>Let S be a stopping time and A belong to $\underline{\underline{F}}_S$. Then we denote by S_A the following stopping time</u> :

$$S_A(\omega)= S(\omega) \text{ if } \omega\in A \quad , \quad S_A(\omega)=+\infty \text{ if } \omega\in A^c .$$

26. COROLLARY (to theorem 24). <u>The well-measurable σ-field is also generated, up to evanescent sets, by adapted processes whose sample functions are right continuous on</u> $[0,\infty[$, <u>and possess a finite left hand limit at every point of</u> $]0,\infty[$

PROOF. The indicators of intervals $[S,T[$ are processes of this kind.

---

[1] Because of right continuity of X and the strict inequality ($>\varepsilon$), the full strength of 14, requiring capacity theory, isn't needed.

27. A nice similar remark of DELLACHERIE : the σ-field $\mathcal{P}$ is generated by adapted continuous processes (with the obvious consequence that $\mathcal{P} \subset \mathcal{W}$, a result which can also be easily proved by other means) . To check it, remark first that, for any stopping time S, the process $X_t=(t-S)^+$ is adapted and sample continuous, and $]S,\infty[= \{X>0\}$. Thus the σ-field generated by continuous adapted processes contains all sets $]u,v]\times H$ ($H\in\underline{\underline{F}}_u$). It also contains $\mathbb{R}_+\times H$ ($H\in\underline{\underline{F}}_0$), since the indicator of this set is an adapted continuous process, and $]0,\infty[\times \Omega$; taking differences we get that it contains $\{0\}\times H$, and the remark is proved.

Another technically useful result on $\mathcal{P}$ is the following one :

28. THEOREM. Let $\mathcal{P}_n$ be the σ-field consisting of all sets

(28.1) $\bigcup_k [k.2^{-n},(k+1)2^{-n}[\times A_k$, where $A_0\in\underline{\underline{F}}_0$, $A_k\in\underline{\underline{F}}_{(k-1)2^{-n}}$ for $n\geq 1$.

Then $\mathcal{P}_n\subset\mathcal{P}_{n+1}\subset\mathcal{P}$ for all n, and $\bigcup_n \mathcal{P}_n$ generates $\mathcal{P}$.

PROOF. Remark first the meaning of this result : $\mathcal{P}_n$ "is" the predictable σ-field of the discrete family $\underline{\underline{G}}_k=\underline{\underline{F}}_{k.2^{-n}}$.

The indicator of $[k.2^{-n},(k+1)2^{-n}[\times A_k$ is the limit as $\varepsilon\to 0$ of that of $]\, k.2^{-n}-\varepsilon,\ (k+1)2^{-n}-\varepsilon]$, for $k>0$ (the case k=0 is left to the reader !). The inclusion $\mathcal{P}_n\subset\mathcal{P}$ follows at once, and $\mathcal{P}_n\subset\mathcal{P}_{n+1}$ is obvious. Set $\mathcal{P}'=\bigvee_n \mathcal{P}_n$; if $A\in\underline{\underline{F}}_0$, $[0,2^{-n}[\times A$ belongs to $\mathcal{P}_n$, and hence $\{0\}\times A$ to $\mathcal{P}'$. Consider $u<v$, $H\in\underline{\underline{F}}_u$, and in (28.1) take $A_k=H$ if $u+2^{-n} \leq k.2^{-n}\leq v+2^{-n}$, $A_k=\emptyset$ otherwise ; letting $n\to\infty$ it comes that $]u,v]\times H \in \mathcal{P}'$, and the theorem is proved.

PREDICTABLE STOPPING TIMES

29. DEFINITION. Let T be a stopping time, and (T_n) be a sequence of stopping times. We say that the sequence (T_n) foretells T[1] if T_n increases with n , $\lim_n T_n=T$, and $T_n<T$ for all n on $\{T>0\}$. If there exists a sequence foretelling T, the stopping time T is said to be predictable.

Since any r.v. variable a.s. equal to a stopping time still is a stopping time, this definition wouldn't be weakened by the insertion of " a.s." at all places. In particular, any stopping time which is a.s. equal to a predictable one is predictable. But it is convenient to define foretelling without exceptional sets. The most trivial example of predictable stopping times is $T+\varepsilon$, where T is

1. If these properties hold true not on the whole of Ω, but on a set A, it is convenient to say that (T_n) foretells T on A.

14

any stopping time, and $\varepsilon>0$: a foretelling sequence is $T_n = (T+\varepsilon-\frac{1}{n})^+\wedge n$.

Typical examples of non-predictable stopping times are heard by a Geiger counter. However, unpredictability in the above mathematical sense is compatible with a fair amount of practical foretelling. Think for instance of earthquakes : one may know that energy is slowly accumulating in faults, and gather from small seisms precious information concerning the nearness of a major catastrophe. But we never can reach the absolute certainty concerning its occurence which is demanded in definition 29.

The discrete analogue of definition 29 is the following : with respect to a discrete family $(\underline{\underline{F}}_n)$, a stopping time T is said to be predictable if for all $n\geq 1$ $\{T\leq n\}$ belongs to $\underline{\underline{F}}_{n-1}$.

We give now some elementary properties of predictable stopping times .

30. THEOREM. If S and T are predictable stopping times, so are $S\vee T$ and $S\wedge T$. If (S_n) is a sequence of predictable stopping times, $\sup_n S_n$ is predictable. If the sequence (S_n) is stationary (i.e. if for every ω there is an integer N such that $S_{N+n}(\omega)=S_N(\omega)$ for every n), then $\inf_n S_n$ is predictable.

PROOF. If (S_n), (T_n) foretell respectively S and T, then$(S_n\wedge T_n)$, $(S_n\vee T_n)$ foretell $S\wedge T$, $S\vee T$. For the second property, we may assume that S_n increases with n. Let (S_{nm}) foretell S_n : then the sequence $S'_m = S_{1m}\vee S_{2m}\vee\ldots\vee S_{mm}$ foretells $\sup_n S_n$. For the third property, we may assume that S_n decreases. Denote by d(x,y) the distance $|e^{-x}-e^{-y}|$ on $\overline{\mathbb{R}}_+$, by(S_{nm}) a sequence of stopping times foretelling S_n, and chose for each n a m_n so large that $P\{d(S_{nm_n},S_n)>\frac{1}{n}\}\leq 2^{-n}$. Finally, set $S'_k = \inf_{n\geq k} S_{nm_n}$. We are going to prove that the sequence (S'_k) foretells $S'=\inf_n S_n$ on a set of full measure - since random variables a.s. equal to stopping times are stopping times, this will imply the desired result.

It is obvious that S'_k increases with k, that $S'_k\leq S'$. Since the sequence (S_n) is stationary, the property $S_{nm_n} < S_n$ on $\{S_n>0\}$ implies $S'_k<S'$ on $\{S'>0\}$. Finally, the BOREL-CANTELLI lemma (trivial part !) implies that for a.e. ω we have $d(S_{nm_n}(\omega),S_n(\omega)) \leq 1/n$ for all large enough n, therefore $d(S'_k(\omega),S'(\omega))\rightarrow 0$ and the theorem follows.

THE SECTION THEOREM

We are going to prove now a very useful result. The proof given in [1] was awful, and didn't concern explicitly the predictable case. It was much simplified by CORNEA-LICEA [6], and then almost trivialized by DELLACHERIE, whose proof is given here .

31. We use the following notation : π is the projection mapping from $\mathbb{R}_+\times\Omega$ onto Ω. If A is any subset of Ω, we denote by $P^*(A)$ the outer probability of A, i.e. $\inf_{\substack{B\supset A\\ B\in\underline{\underline{F}}}} P(B)$, and if M is any subset of $\mathbb{R}_+\times\Omega$, we set $P^\times(M)=P^*(\pi(M))$. For any random variable U with value in $\overline{\mathbb{R}}_+$ (not necessarily a stopping time) we denote by [U] the "graph" of U, that is

(31.1) $\{ (s,\omega) : s=U(\omega)<\infty \}$

Since [U] is the inverse image of $\{(x,y)\in\overline{\mathbb{R}}_+\times\overline{\mathbb{R}}_+ : x=y<\infty\}$ by the measurable mapping $(s,\omega)\mapsto(s,U(\omega))$, [U] is measurable. We need the following result from measure theory :

Let A belong to $\underline{\underline{B}}(\mathbb{R}_+)\times\underline{\underline{F}}$, and let ε be a number >0. There then exists a random variable U with values in $\overline{\mathbb{R}}_+$, such that

(31.2) $[U]\subset A$, $P^\times([U]) \geqq P^\times(A) - \varepsilon/2$.

Note that $P^\times([U])=P\{U<\infty\}$. This is a rather easy consequence of the capacity theorem, like in n°14, but we aren't going to prove it. See [5][1], which contains a much fuller account of this theory, or the next edition of [1].

32. THEOREM. Let A be a well-measurable set, and ε be a number >0. There exists a stopping time T such that

(32.1) $[T]\subset A$, $P^\times([T]) \geq P^\times(A)-\varepsilon$.

If A is a predictable set, there exists a predictable T with these properties.

PROOF. We begin with the predictable case, which is more important for us. We use the random variable U from (31.2) to define a bounded, positive measure μ on $\mathbb{R}_+\times\Omega$ as follows : for any measurable function (process) X on $\mathbb{R}_+\times\Omega$, which is either positive or bounded

$$\mu(X) = E[X(U(\omega),\omega)I_{\{U<\infty\}}(\omega)]$$

The mass of μ is greater than $P^\times(A)-\varepsilon/2$ according to (31.2). Consider the σ-fields $\mathcal{P}_n$ of n°28. Since A is $\mathcal{P}$-measurable, the theorem

1 Chapter I, theorem 37.

16

of CARATHEODORY implies the existence of sets $C_n \in \mathcal{P}_n$ such that

$$C_{n+1} \subset C_n \text{ for all } n \ , \ \bigcap_n C_n = C \subset A \ , \ \mu(C) \geqq \mu(A) - \varepsilon/2$$

(no capacity theory is used here, just classical measure theory !)
Set

$$T_n(\omega) = \inf \{ t : (t,\omega) \in C_n \}$$

$$T(\omega) = \inf \{ t : (t,\omega) \in C \}$$

T_n is a dyadic valued stopping time, and $\{T_n = k2^{-n}\} \in \underline{\underline{F}}_{(k-1)2^{-n}}$ if $k \geqq 1$. Hence $(T_n - t)^+$ is a stopping time for $t \leqq 2^{-n}$, and T_n is predictable. On the other hand, the sets C_n are " closed from the right"[1]; and the same is true for C : one deduces easily from it that $T_n \uparrow T$ (whence the predictability of T), and that $[T] \subset C \subset A$. Since μ is the "unit mass" on $[U]$, the fact that $\mu(C) \geqq \mu(A) - \varepsilon/2 \geq P^X(A) - \varepsilon$ implies $P^X(C) \geqq P^X(A) - \varepsilon$, and the proof ends with the remark that $P^X(C) = P\{T < \infty\}$.

The proof for the well-measurable case is similar, but somewhat simpler, the natural generators for the σ-field $\mathcal{W}$ being already given as sets closed from the right. Let us say that a well-measurable set is elementary if its indicator function is a right continuous process with left hand limits. It is obvious that elementary sets constitute an algebra, which generates $\mathcal{W}$ up to evanescent sets (n° 24 : intervals $[S,T[$ are elementary sets). One then applies the same reasoning as above to this algebra instead of $\bigcup_n \mathcal{P}_n$.

Let us give just one application of theorem 32 :

33. THEOREM. Let U be a positive random variable. Then U is a stopping time (a predictable stopping time) if and only if $[U]$ is a well-measurable set (a predictable set).

PROOF. If U is a stopping time, $[U] = \bigcap_n [U, U+\frac{1}{n}[$ is well measurable. If U is predictable, foretold by a sequence (U_n), then $[U] = \{0\} \times \{U=0\}) \cup (\bigcap_n]U_n, U])$ is a predictable set.

Conversely, assume $[U]$ is well-measurable. According to theorem 34 one may find stopping times U_n such that $[U_n] \subset [U]$, $P\{U_n < \infty\} \geqq$

---

[1] A subset A of $\mathbb{R}$ is closed from the right if the limit of any decreasing sequence of elements of A belongs to A. A subset C of $\mathbb{R}_+ \times \Omega$ is closed from the right if (a.a.) sections $C(\omega)$ are closed from the right in $\mathbb{R}_+$.

$P\{U<\infty\} = 2^{-n}$. Therefore $U_n \to U$ a.s., implying that U is a stopping time.

If $[U]$ is predictable, we use the same reasoning to construct predictable stopping times U_n with the above properties. The sequence (U_n) is a.s. stationary, and $U = \inf_n U_n$ a.s.. Then we apply theorem 30.

THE σ-FIELD $\underline{\underline{F}}_{T-}$

We have announced in n° 10 that we would associate to any stopping T, besides the usual σ-field $\underline{\underline{F}}_T$, another σ-field describing the past " just before T" . The definition of this σ-field $\underline{\underline{F}}_{T-}$ is due to CHUNG and DOOB [7]. Note that $\underline{\underline{F}}_{T-}$ is defined through generators, while $\underline{\underline{F}}_T$ wasn't.

34. DEFINITION. Let T be a stopping time. We denote by $\underline{\underline{F}}_{T-}$ the σ-field generated by $\underline{\underline{F}}_0$ and the sets $A\cap\{t<T\}$ $(t\in\mathbb{R}_+ , A\in\underline{\underline{F}}_t)$

All generators of $\underline{\underline{F}}_{T-}$ belong to $\underline{\underline{F}}_T$, hence $\underline{\underline{F}}_{T-}\subset\underline{\underline{F}}_T$. Also T is $\underline{\underline{F}}_{T-}$-measurable (take $A=\Omega$ in the definition). Note that $\underline{\underline{F}}_{T-}\subset\underline{\underline{F}}_\infty$ and that, for any $A\in\underline{\underline{F}}_\infty$, $A\cap\{T=\infty\}\in\underline{\underline{F}}_{T-}$ (it suffices to prove it for $A\in\underline{\underline{F}}_n$, and this set is $\bigcap_{m>n} A\cap\{m<T\}$).

The following result shows how the pair $\underline{\underline{F}}_{T-},\underline{\underline{F}}_T$ corresponds to the pair predictable/well-measurable.

35. THEOREM. Let η be a $\underline{\underline{F}}_\infty$-measurable random variable. Then η is $\underline{\underline{F}}_T$-measurable ($\underline{\underline{F}}_{T-}$-measurable) if and only if there exists a well-measurable (predictable) process Y such that $\eta=Y_T$ on $\{T<\infty\}$.

PROOF. Let f be the mapping $\omega\mapsto(T(\omega),\omega)$ from $\{T<\infty\}$ to $\mathbb{R}_+\times\Omega$. Since $\underline{\underline{F}}_\infty,\underline{\underline{F}}_{T-}$ and $\underline{\underline{F}}_T$ induce the same σ-field on $\{T=\infty\}$, the theorem amounts to the following : the σ-fields induced by $\underline{\underline{F}}_T$, $\underline{\underline{F}}_{T-}$ on $\{T<\infty\}$ are exactly $f^{-1}(\mathcal{W})$, $f^{-1}(\mathcal{P})$ (see [1], chapter I, theorem 18). For the predictable case, we choose the generators of $\mathcal{P}$ from n°24 : $\{0\}\times A$ $(A\in\underline{\underline{F}}_0)$ and $]t,\infty]\times A$ $(A\in\underline{\underline{F}}_t)$, and the property becomes obvious. For the well-measurable case, we generate $\mathcal{W}$ through sets $[S,\infty[$, S denoting a stopping time, and leave the details to the reader.

36. We give now some simple properties of $\underline{\underline{F}}_{T-}$. For complete proofs see [5], chapter III, §3 . S,T denote stopping times.

a) If $S\leq T$, $\underline{\underline{F}}_{S-}$ is contained in $\underline{\underline{F}}_{T-}$.
Hint : if $A\in\underline{\underline{F}}_t$, $A\cap\{t<S\} = (A\cap\{t<S\})\cap\{t<T\}$, and $A\cap\{t<S\}\in\underline{\underline{F}}_t$.

b) If $A\in\underline{\underline{F}}_S$, $A\cap\{S<T\}$ belongs to $\underline{\underline{F}}_{T-}$.
Hint : $A\cap\{S<T\} = \bigcup_r (A\cap\{S\leq r\})\cap\{r<T\}$, r rational .

c) If $S\leq T$, $S<T$ on $\{0<T<\infty\}$, then $\underline{\underline{F}}_S\subset\underline{\underline{F}}_{T-}$.
Hint : If $A\in\underline{\underline{F}}_S$, $A=(A\cap\{T=\infty\})\cup[\bigcup_r(A\cap\{S\leq r\})\cap\{r<T\}]\cup(A\cap\{S=0\})$

d) If stopping times T_n increase to T, then $\underline{\underline{F}}_{T-}=\bigvee_n \underline{\underline{F}}_{T_n-}$.
Hint : $\supset$ obvious, to prove $\subset$ look at the generators.

e) If T is predictable, and the sequence (T_n) foretells T, then $\underline{\underline{F}}_{T-}=\bigvee_n \underline{\underline{F}}_{T_n}$.
Hint : This last σ-field contains $\underline{\underline{F}}_{T-}$ from d), and is contained in it from c).

The following theorem is a little less obvious.

37. THEOREM. Let T be a predictable stopping time. Then $A\in\underline{\underline{F}}_\infty$ belongs to $\underline{\underline{F}}_{T-}$ if and only if T_A (n°25) is predictable.

PROOF. If $A\in\underline{\underline{F}}_{T-}$, let X be a predictable process such that $I_A=X_T$ on $\{T<\infty\}$. Then $[T_A]=[T]\cap\{X=1\}$ is predictable, and T_A is a predictable stopping time according to n° 33. Conversely, if T_A is predictable, $Y=I_{[T_A]}$ is a predictable process such that $I_A=Y_T$ on $\{T<\infty\}$.

INCREASING PROCESSES

38. We are going to introduce here the basic (though mostly trivial) lemmas for the theories of DOOB's decomposition in continuous time , and of stochastic integrals.

We call increasing process any finite valued process A , adapted and right continuous, such that $A_0=0$, and the sample functions of which are increasing. We shall occasionally stumble upon "non adapted increasing processes" , i.e. relative to the family $\underline{\underline{F}}'_t=\underline{\underline{F}}$. If A is an increasing process, we shall generally denote by A_∞ the limit of A_t as $t\to\infty$, though the correct notation would be $A_{\infty-}$. When the notation A_∞ appears without a special explanation, it must be so understood. An increasing process A is said to be <u>integrable</u> if $E[A_\infty]<\infty$, <u>locally integrable</u> if there exists an increasing sequence (T_n) of stopping times, such that $T_n \to \infty$ a.s. and $E[A_{T_n}] < \infty$ for every n .

We denote by $\underline{\underline{V}}$ the linear space consisting of <u>differences of increasing processes</u> , and call its elements <u>processes of finite variation</u>. Differences of integrable (locally integrable) increasing processes are called processes of integrable (locally integrable) variation, and the corresponding spaces are denoted by $\underline{\underline{I}}$ and $\underline{\underline{LI}}$. In the discrete case, all processes belong to $\underline{\underline{V}}$.

Let X be an adapted , right **continuous** process, such that $X_0=0$ and that every path $X_\cdot(\omega)$ is a function of finite variation on every finite interval $[0,t]$. Then X belongs to $\underline{\underline{V}}$, and can be canonically represented as a difference of increasing processes. First of all, the process X has left limits on $]0,\infty[$, the process X_t-X_{t-} is well-measurable, and therefore the set

$$J_\varepsilon = \{(t,\omega) : |X_t(\omega)-X_{t-}(\omega)| \geqq \varepsilon \} \qquad (\varepsilon>0)$$

is a well-measurable set whose sections $J(\omega)$ are discrete. Considering the hitting time T_1 of J_ε , then the hitting time T_2 of $J_\varepsilon \setminus [T_1]$... we represent J_ε as the union of a sequence of disjoint graphs of stopping times. It is then obvious to check that the processes

(38.1)
$$\sum_{\substack{s\in J_\varepsilon(\omega)\\ s\leqq t}} X_s(\omega)-X_{s-}(\omega)$$
$$\sum_{\substack{s\in J_\varepsilon(\omega)\\ s\leqq t}} |X_s(\omega)-X_{s-}(\omega)|$$

are well-measurable. As $\varepsilon\to 0$, the first one tend to the <u>jump part</u> of X

(38.2) $$X_t^{(j)} = \sum_{s\leqq t} X_s - X_{s-}$$

and the second one to the total variation process of $X^{(j)}$

(38.3) $$\sum_{s\leqq t} |X_s - X_{s-}|$$

20

The process $X^{(c)}=X-X^{(j)}$ is the continuous part of X : its total variation $\int_0^t |dX_s^{(c)}(\omega)|$ can be computed as

$$\lim_{n\to\infty} \sum_{k=1}^{2^n} |X_{kt/2^n}(\omega)-X_{(k-1)t/2^n}(\omega)|$$

and therefore is an adapted, continuous increasing process. Putting together these results, we find that the total variation process for X, which we shall denote by $\{X\}$

$$(38.4) \qquad \{X\}_t = \int_0^t |dX_s|$$

is an increasing process. It is then obvious that X belongs to $\underline{\underline{V}}$, with the decomposition $\frac{1}{2}(\{X\}+X) - \frac{1}{2}(\{X\}-X)$. If X belongs to $\underline{\underline{I}}$ ($\underline{\underline{LI}}$) these processes are integrable (locally integrable).

STOCHASTIC STIELTJES INTEGRALS

39.We are going to develop here the trivial part of the stochastic integration theory : integration with respect to processes of finite variation. It may help the reader to be recalled at this step that, later on, we shall integrate with respect to martingales (i.e. processes which, typically, aren't of finite variation), and still later mix the two kinds of integrals into a unified theory .

Let A be a process of finite variation, and let V be a measurable process. Assume that for every t

$$(39.1) \qquad \int_0^t |V_s(\omega)||dA_s(\omega)| < \infty \quad \text{a.s.}$$

(integration with respect to the absolute value of the measure dA_s , that is, with respect to $d\{A\}_s$). Throwing away a set of probability 0, we may assume that (39.1) holds true for all t and ω. Then, according to FUBINI's theorem, the process

$$(39.2) \qquad W_t(\omega) = \int_0^t V_s(\omega)dA_s(\omega)$$

is right continuous and has finite variation. If V is well-measurable, then W is also adapted, and thus belongs to $\underline{\underline{V}}$. If V is predictable, we shall call W the transform of A by V, as in the discrete case, and denote it by V.A .

We shall not attempt to build a detailed theory of stochastic STIELTJES integrals : this chapter already is soporific enough. We just want to prove two useful lemmas (taken from [1]), with some comments.

40. THEOREM. Let A be an increasing process. Let X and Y be two positive measurable processes, such that for every stopping time T

(40.1) $$E[X_T I_{\{T<\infty\}}] = E[Y_T I_{\{T<\infty\}}]$$

Then we have

(40.2) $$E[\int_0^\infty X_s dA_s] = E[\int_0^\infty Y_s dA_s]$$

PROOF (Sketchy : for details see [1], VII.T15 ou [5], IV.T46).
Set for every $t \geqq 0$

$$c_t = \inf \{ s : A_s > t \}$$

Then c_t is a stopping time, the function $c_{\cdot}(\omega)$ is increasing and right continuous (with the value $+\infty$ allowed) and we have

(40.3) $$E[\int_0^\infty X_s dA_s] = E[\int_0^\infty X_{c_t} I_{\{c_t<\infty\}} dt]$$

([1], VII.T12 or [5], IV.T44), and the similar equality with Y instead of X . From (40.1) we get that $E[X_{c_t} I_{\{c_t<\infty\}}]=E[Y_{c_t} I_{\{c_t<\infty\}}]$ and an application of FUBINI's theorem concludes the proof.

41. REMARKS. a) The theorem can be extended to slightly different situations. For instance, if X and Y are bounded, (40.2) will be true if A has integrable variation, without being increasing. It would be pointless to write explicitly such trivial generalizations.

b) Let X and Y be positive or bounded. If (40.1) is true for all T, then for a given T it will hold for all T_B ,$B\in\underline{\underline{F}}_T$ (n°25), and therefore (40.1) is equivalent to the apparently stronger property

(41.1) $$E[X_T I_{\{T<\infty\}} | \underline{\underline{F}}_T] = E[Y_T I_{\{T<\infty\}} | \underline{\underline{F}}_T]$$

c) (40.2) can be similarily strengthened. Let T be a stopping time. Define $\Omega'=\{T<\infty\}$ with the induced probability law, $\underline{\underline{F}}'_t=\underline{\underline{F}}_{T+t}$, $X'_t=X_{T+t}$, $Y'_t=Y_{T+t}$, $A'_t=A_{T+t}-A_T$. If S is a stopping time of the family $(\underline{\underline{F}}'_t)$, T+S (defined as $+\infty$ out of Ω') is a stopping time of $(\underline{\underline{F}}_t)$, and (40.2) on this space gives us

(41.2) $$E[\int_T^\infty X_s dA_s] = E[\int_T^\infty Y_s dA_s]$$

Considering again T_B $(B\in\underline{\underline{F}}_T)$ instead of T, we may replace the expectations by conditional expectations relative to $\underline{\underline{F}}_T$ as in (41.1).

22

d) By difference between (40.2) and (41.2), we have formally

(41.3) $$E[\int_0^T X_s dA_s] = E[\int_0^T Y_s dA_s]$$

It may be interesting to know that (41.3) may be proved without taking differences, just by applying (40.2) to the increasing process $A_{t\wedge T}$. Therefore (41.3) holds without integrability restrictions : A increasing, X and Y positive and (40.1) suffice.

We prove now a slightly different version of theorem 40 (it may be omitted at the first reading).

42. THEOREM. Let A be a predictable increasing process. Let X and Y be two positive measurable processes, such that for any predictable stopping time T (40.1) is true. Then (40.2), (41.2), and (41.3) still are true (for any stopping time T, predictable or not).

PROOF. Instead of c_t , consider the stopping times

$$d_t = \inf \{ s : A_s \geq t \}$$

d_t is the hitting time of $G_t = \{ (s,\omega) : A_s(\omega) \geq t \}$. We have $d_t(\omega) \neq c_t(\omega)$ if and only if there are two instants $s_1 < s_2$ such that $A_{s_1}(\omega) = A_{s_2}(\omega) = t$, and therefore if $d_t(\omega) \neq c_t(\omega)$ there exists a rational r such that $A_r(\omega) = t$. One deduces easily from it that the set of all t such that $P\{d_t \neq c_t\} \neq 0$ is at most countable, and therefore (40.3) implies

$$E[\int_0^\infty X_s dA_s] = E[\int_0^\infty X_{d_t} I_{\{d_t < \infty\}} dt]$$

We have the same equality with Y instead of X, and we can conclude as above provided we show that d_t is predictable. This is easy : since A is a predictable process, $G_t = \{(s,\omega) : A_s(\omega) \geq t \}$ is a predictable set. Also, A being right continuous, we have $A_{d_t} \geq t$, that is $[d_t] \subset G_t$. Hence we may write $[d_t] = [0, d_t] \cap G_t$; the first set on the right is predictable , since its indicator is a left continuous adapted process. The intersection of two predictable sets being predictable, $[d_t]$ is predictable, and we apply theorem 33.

Our second important lemma is " dual" to the first one[1]. Instead of two processes and a measure (increasing process) we have one process and two measures.

[1]For details, see [5].

43 . THEOREM. Let A and B be two increasing processes (not necessarily adapted) such that A_t and B_t are integrable for each t, and for s<t

(43.1) $$E[A_t - A_s | \underline{\underline{F}}_s] = E[B_t - B_s | \underline{\underline{F}}_s]$$

Let X be a positive predictable process. Then

(43.2) $$E[\int_0^\infty X_s dA_s] = E[\int_0^\infty X_s dB_s] .$$

PROOF. It is sufficient to deal with the case where A_∞ and B_∞ are integrable (stop both processes at n, then let $n \to \infty$), and X is bounded. Using a monotone class argument, it suffices to prove the theorem for indicators of sets belonging to $\mathcal{P}_n$ (n°28), and in this case (43.2) is a trivial consequence of (43.1).

44. REMARKS. a) If X is bounded, the result extends at once to the case of two processes A,B of integrable variation satisfying to (43.1).

b) If T is a stopping time, the indicator of $]T,\infty[$ is predictable. Applying (43.2) to $XI_{]T,\infty[}$ instead of X, we get

(44.1) $$E[\int_T^\infty X_s dA_s] = E[\int_T^\infty X_s dB_s]$$

and even, applying this to T_B ($B \in \underline{\underline{F}}_T$)

(44.2) $$E[\int_T^\infty X_s dA_s | \underline{\underline{F}}_T] = E[\int_T^\infty X_s dB_s | \underline{\underline{F}}_T] .$$

45. Let us say that two processes A and B, both right continuous and such that A_t and B_t are integrable, are associated if (43.1) holds. This corresponds exactly to the discrete definition given in n°18, and we shall use the notation $A \sim B$ as in the discrete case. If A and B are adapted, $A \sim B$ means that A-B is a martingale. Then theorem 43 can be stated as follows : assume tha A,B are associated and have integrable variation, and that X is bounded and predictable . Then X.A and X.B also are associated. This property wouldn't be true if " predictable" was replaced by " well-measurable" , and is the basic explanation of the role played by predictable processes in the theory of stochastic integrals.

CHAPTER II . DISCRETE MARTINGALE THEORY

§ 1 . The classical theory : Sampling, Inequalities, Convergence

This chapter doesn't pretend to describe all the facets of discrete martingale theory : many beautiful and modern topics, like optimal stopping , are omitted. On the other hand, it contains probably more than will be needed for the extension to the continuous case, and the theory of stochastic integrals. Between the first and second drafts, the author has benefited from a copy of a set of lectures by NEVEU (Paris, 1970-71), and from conversations with S.D. CHATTERJI.

All processes in this chapter are defined on a probability space $(\Omega,\underline{\underline{F}},P)$, assumed to be complete. The words martingale, stopping time, predictable... all refer to the same fixed family $(\underline{\underline{F}}_n)_{n\in\mathbb{N}}$, unless the contrary is specifically stated.

DOOB'S OPTIONAL SAMPLING THEOREM (BOUNDED STOPPING TIMES)

1. THEOREM Let X be a martingale (supermartingale) and V a precess which is predictable (predictable $\geqq 0$). If the random variables $(V.X)_n$ are integrable, the transformed process V.X is a martingale (supermartingale).

PROOF. $E[(V.X)_{n+1}-(V.X)_n|\underline{\underline{F}}_n] = E[V_{n+1}(X_{n+1}-X_n)|\underline{\underline{F}}_n]=V_{n+1}E[X_{n+1}-X_n|\underline{\underline{F}}_n]$ since V_{n+1} is $\underline{\underline{F}}_n$-measurable. If X is a martingale, this last conditional expectation is 0, and V.X is a martingale. If X is a supermartingale, this expectation is $\leqq 0$, and so is $V_{n+1}E[X_{n+1}-X_n|\underline{\underline{F}}_n]$ since V is $\geqq 0$. Therefore V.X is a supermartingale.

2. COROLLARY. Let X be a martingale (supermartingale), and T be a stopping time. Then the stopped process X^T is a martingale (supermartingale).

PROOF. In the preceding theorem take $V_n=I_{\{n\leq T\}}$, so that $X^T_n=(V.X)_n$. Integrability is obvious : $|X^T_n|\leqq |X_0|+...+|X_n|$.

The intuitive application of theorem 1 is the following : think of X_n as of the fortune of a gambler just after the n-th play. Then $(V.X)_n$ may be considered as the fortune of a second gambler who is betting on the success of the first one. The martingale property of X

$$E[X_{n+1}-X_n|\underline{\underline{F}}_n] = 0$$

is called fairness in gambling theory : the idea that a gamble is " fair " if the expected return is 0 is among the oldest ones in probability theory, going back to PASCAL at least. Similarily, the supermartingale property means subfairness (from the gambler's, not the casino's, point of view). Thus theorem 1 means that " betting on a fair gambler is fair gambling" .

The following basic result is called DOOB's optional sampling theorem . The boundedness assumption on the stopping times will be weakened later on (n° 28-29).

3. THEOREM. *Let X be a supermartingale (martingale), and let S and T be two bounded stopping times such that $S\leq T$. Then X_S and X_T are integrable and we have*

(3.1) $$X_S \geq E[X_T|\underline{\underline{F}}_S] \quad \text{a.s. (= in the martingale case).}$$

PROOF. The martingale case follows from the supermartingale case applied to X and -X. Set $V_n=I_{\{S<n\leq T\}}$, a positive and predictable process , and Y=V.X . Then Y is a supermartingale, $Y_0=0$, and if k dominates T we have $Y_k=X_T-X_S$. The supermartingale property of Y then gives $0\geq E[Y_k]=E[X_T-X_S]$. To get the conditional form (3.1) choose $A\in\underline{\underline{F}}_S$, and apply the preceding reasoning to $S'=S_A\wedge k$,$T'=T_A\wedge k$. The desired inequality $\int_A(X_T-X_S)P \leq 0$ comes out.

THE MAXIMAL INEQUALITY

4. DEFINITION. *Let X be any process. Then we set* $\|X\|_p = \sup_n \|X_n\|_p$ $(1\leq p\leq\infty)$. *A process X such that $\|X\|_p<\infty$ is said to be bounded in L^p.*

We are going to state only the most useful form of the basic maximal inequality for supermartingales. But the proof gives sharper inequalities, among which (5.6) is specially interesting.

5. THEOREM. *Let X be a supermartingale and c be >0. Then*

(5.1) $$P\{ \sup_n |X_n| \geq c \} \leq A\frac{\|X\|_1}{c}$$

(A *can be taken equal to 1 if $X\geq 0$, $X\leq 0$, or X is a martingale. In the general case, one may take* A=3).

(It suffices to prove the inequality with " >c " instead of "$\geq$c").

PROOF. Let us choose some k and set

(5.2) $T(\omega) = \inf \{ n : n \leqq k , X_n(\omega) \geqq c \}$

If no such n exists, we set $T(\omega)=k$. Then T is a stopping time $\leqq k$, to which theorem 3 can be applied. Therefore we have $E[X_0] \geqq E[X_T]$.

If $X_n(\omega) \geqq c$ for some $n \leqq k$, we have $X_T \geqq c$. Otherwise, $X_T = X_k$. Accordingly, we have

(5.3) $E[X_0] \geqq cP \{ \sup_{n \leqq k} X_n \geqq c \} + \int_{\{ \sup_{n \leqq k} X_n < c \}} X_k P$

Letting $k \to \infty$

(5.4) $cP\{ \sup_n X_n \geqq c \} \leqq E[X_0] + \sup_k E[X_k^-] \leqq 2\|X\|_1$

Note that (5.1) is proved (with $\lambda=1$) when $X^- = 0$, i.e. $X \geqq 0$! To deal with negative values $\leqq -c$, set

(5.5) $T(\omega) = \inf \{ n : n \leqq k , X_n(\omega) \leqq -c \}$

$T(\omega)=k$ if no such n exists. We apply theorem 3 to get $E[X_T] \geqq E[X_k]$ which implies , as above

$$E[X_k] \leqq E[X_T] \leqq -cP\{ \inf_{n \leqq k} X_n \leqq -c \} + \int_{\{ \inf_{n \leqq k} X_n > -c \}} X_k P$$

or

(5.6) $cP\{ \inf_{n \leqq k} X_n \leqq -c\} \leqq \int_{\{ \inf_{n \leqq k} X_n \leqq -c\}} -X_k P \leqq E[X_k^-] \leqq \|X\|_1$

Letting $k \to \infty$ we get (5.1) (with $\lambda=1$) for the case $X \leqq 0$, and also for the martingale case - since then $-|X|$ is a negative supermartingale from I.8. Adding (5.4) and (5.6) we find in the general case

(5.7) $cP\{ \sup_n |X_n| \geqq c \} \leqq E[X_0] + 2.\sup_k E[X_k^-] \leqq 3.\|X\|_1$.

6. We leave for a short while the main stream, to indicate some sharpenings of the above inequalities. This n° should be omitted at a first reading.

We start with the remark that, if (X_n) is a supermartingale , and if we define the law P^A on Ω by

$$P^A(B) = \frac{P(A \cap B)}{P(A)} \qquad (A \in \underline{\underline{F}}_0 , P(A)>0 \ ; B \in \underline{\underline{F}})$$

then X still is a supermartingale for P^A. To shorten notations, set $X^* = \sup_n X_n$, and apply (5.4) to P^A instead of P. One gets

$$cP\{ X^* \geqq c , A \} \leqq \int_A X_0 P + \sup_k \int_A X_k^- P$$

which may be written as the conditional version of (5.4)

(6.1) $cP\{X^* \geqq c \mid \underline{\underline{F}}_0\} \leqq X_0 + \sup_k E[X_k^- \mid \underline{\underline{F}}_0]$

Denote now by Λ some positive, $\underline{\underline{F}}_0$-measurable random variable. Assume first that for some $\varepsilon>0$ we have $\Lambda \geqq \varepsilon$. The process $(\frac{X_n}{\Lambda})$ is a supermartingale. Applying to it (6.1) with c=1, we get

(6.2) $\Lambda P\{X^* \geqq \Lambda \mid \underline{\underline{F}}_0\} \leqq X_0 + \sup_k E[X_k^- \mid \underline{\underline{F}}_0]$

This inequality is true without the assumption $\Lambda \geqq \varepsilon$: take $A_0 = \{\Lambda = 0\}$ $A_n = \{\Lambda \geqq 1/n\}$ for n>0 . These sets belong to $\underline{\underline{F}}_0$, and (6.2) holds on each of them - trivially on A_0, and for n>0 according to the above particular case applied to the law P^{A_n}.

Inequality (6.2) isn't particularly useful, but the idea of getting conditional forms of inequalities does have applications .

BOUNDS IN L^p FOR MARTINGALES

7 Consider a mapping $H \mapsto H^*$ from the space of elementary r.v. on $(\Omega, \underline{\underline{F}})$ to the space of a.s. finite r.v., such that

$$(tH)^* = tH^* \text{ for } t \geqq 0 \ , \ (H+K)^* \leqq H^* + K^*$$

We say that the mapping if <u>of type (r,r)</u> $(1 \leqq r \leqq \infty)$ if $\|H^*\|_r \leqq A_r \|H\|_r$ (H elementary), <u>of weak type (1,1)</u> if $cP\{|H^*|>c\} \leqq A\|H\|_1$ ($c>0$, H elementary). Then a particular case of the celebrated MARCINKIEWICZ interpolation theorem (see for instance STEIN [8], p.21 and 272) asserts that

If the mapping is of weak type (1,1) and of type (r,r), then it is also of type (p,p) for $1<p<r$, and A_p depends only on A, A_r, r .

For instance, denote by H^* the r.v. $\sup_n |E[H|\underline{\underline{F}}_n]|$: this mapping is trivially of type (∞,∞), and (5.1) means exactly that it is of weak type (1,1). According to the MARCINKIEWICZ theorem, it is of type (p,p) for all p>1. We are going to prove this result directly (though we shall need the MARCINKIEWICZ theorem later on), borrowing from DOOB's book the elementary and interesting proof it contains, and which is probably due to WIENER.

8. THEOREM. *Let X be a positive submartingale, and set $X^* = \sup_n X_n$. Then we have for p>1*

(8.1) $\| X^* \|_p \leqq q \| X \|_p$

q <u>denoting the exponent conjugate to</u> p ($\frac{1}{p}+\frac{1}{q}=1$) . <u>The theorem applies in particular to</u> $X=|Y|$ <u>for any martingale</u> Y.

PROOF. Set $X_k^* = \sup_{n\leq k} X_n$. It suffices to prove that $\|X_k^*\|_p \leq q\|X_k\|_p$. On the other hand, we have according to (5.6)

$$cP\{X_k^*\geq c\} \leq \int_{\{X_k^*\geq c\}} X_k P$$

and the theorem appears as an obvious consequence of the following lemma.

9. THEOREM. <u>Let</u> U <u>and</u> V <u>be two positive random variables such that</u>

$$(9.1)\qquad cP\{V\geq c\} \leq \int_{\{V\geq c\}} U.P \qquad (c>0)$$

<u>Then we have</u> $\|V\|_p \leq q\|U\|_p$ <u>for</u> $1<p<\infty$ [1].

PROOF. Let F be an increasing continuous function on $[0,\infty[$, with $F(0)=0$. Then setting $L(c)=P\{V\geq c\} \leq \frac{1}{c}\int_{\{V\geq c\}} U.P$

$$E[F\circ V] = -\int_0^\infty F(c)dL(c) \leq \int_0^\infty L(c)dF(c) \quad \text{(integration by parts)}$$

$$(9.2)\qquad \leq \int_0^\infty \frac{dF(c)}{c}\int_{\{V\geq c\}} U.P = E[\, U.\int_0^V \frac{dF(c)}{c}\,] .$$

Let us now take $F(c)=c^p$, we get

$$E[V^p] \leq \frac{p}{p-1}E[U.V^{p-1}] \leq \frac{p}{p-1}\|U\|_p\|V^{p-1}\|_q = \frac{p}{p-1}\|U\|_p(\|V\|_p)^{p/q}$$

(HÖLDER's inequality). Assume first that $\|V\|_p$ is finite, and divide both members by $\|V\|_p^{p/q}$. Since $q=\frac{p}{p-1}$ we get $\|V\|_p\leq q\|U\|_p$, that is, the desired inequality. To remove the finiteness assumtion remark that for every n $V\wedge n$ satisfies (9.1) and the finiteness assumption, and then let $n\to\infty$. This truncation argument, however, fails on an infinite measure space (see below remark b)).

10. REMARKS. a) If one doesn't care for a precise estimate of the constant (that of theorem 9 can be shown to be sharp) the following argument is simpler. Write (9.1) with 2c instead of c, and use the obvious inequalities

$$(10.1)\qquad 2cP\{V\geq 2c\} \leq \int_{\{V\geq 2c\}} U.P \leq \int_{\{U\geq c\}} U.P + \int_{\{U<c,V\geq 2c\}} U.P$$

$$\leq \int_{\{U\geq c\}} U.P + cP\{V\geq 2c\}$$

[1] The inequality is true also for $p=\infty$ (obvious).

Then instead of (9.2) we have

(10.2) $E[F\circ\frac{V}{2}] \leqq E[U\int_0^U \frac{dF(c)}{c}]$

and if $F(c)=c^p$

(10.3) $\|V\|_p \leqq 2q^{1/p}\|U\|_p$

b) This proof works for an infinite measure space too : taking P to be any positive measure, and assuming $\|U\|_p<\infty$ (if it is $+\infty$, there is nothing to prove) we have from (9.1) that $P\{V\geqq 2c\}$ is finite for all $c>0$, and we may take differences in (10.1). Knowing (10.3), we may return to the proof of n°9 and get the sharp inequality on an infinite measure space, without truncation argument.

c) (10.2) is very convenient to deal with functions F which aren't powers of c. For example, take $F(c)=(c-1)^+$. $E[F\circ\frac{V}{2}]$ is greater than $E[\frac{V}{2}]-1$, and we get

(10.4) $E[V] \leqq 2(\ 1+E[U.\log^+ U])$

Here nobody cares for a sharp constant, so we can just as well state

11.THEOREM. Let X be a positive submartingale, and set $X^*= \sup_n X_n$. Then

(11.1) $E[X^*] \leqq 2(\ 1 + \sup_n E[X_n.\log^+ X_n])$

This is a very useful theorem. Though the constant isn't sharp, the theorem is best possible in at least two ways : first, there is no function $\phi(t)$ growing more slowly than $t\log^+ t$, such that $\sup_n E[\phi\circ X_n] < \infty$ would imply $E[X^*]<\infty$ (BLACKWELL and DUBINS [9]). Next, there is a large class of martingales for which the inequality (11.1) is reversible : $\sup_n E[|X_n|\log^+|X_n|] \leqq AE[|X|^*]$ (GUNDY [10])

CONVERGENCE : FIRST PROOF

We shall give two proofs of the basic results on convergence. The shortest way certainly is the classical one through DOOB's upcrossing inequality, which will be given below in n° 23 . However, DOOB's first proof in [11] didn't use upcrossings[1], and proofs without upcrossings are rediscovered periodically. We shall therefore include the nice way to the convergence theorems which was shown to us by CHATTERJI. As we shall see, it gives also quite easily the most useful convergence theorem for martingales with values in a BANACH space.

1 See Appendix 1.

12. DEFINITION. We say that a martingale (X_n) is closed in L^p by a random variable Y if $Y\in L^p$ and $X_n = E[Y|\underline{\underline{F}}_n]$ for every n.

If Y closes (X_n), so does $E[Y|\underline{\underline{F}}_\infty]$.

13. THEOREM. Let X be a martingale, closed in L^p $(1\leq p<\infty)$ by a random variable Y. Then X_n converges a.s. and in the norm of L^p to $E[Y|\underline{\underline{F}}_\infty]$. $X_\infty = \lim_n X_n$ is the only $\underline{\underline{F}}_\infty$-measurable r.v. which closes X in L^1.

PROOF. We start with the case p=1, and the assumption that $\underline{\underline{F}}=\underline{\underline{F}}_\infty$. Denote by $\underline{\underline{H}}$ the set of all random variables $X\in L^1$ such that the martingale $\overline{X}$

(13.1) $$\overline{X}_n = E[X|\underline{\underline{F}}_n]$$

converges to X a.s. and in L^1. $\underline{\underline{H}}$ is a linear space, and contains $L^1(\underline{\underline{F}}_k)$ for every finite k, since then $\overline{X}_n = X$ for $n\geq k$. Now $\bigcup_k L^1(\underline{\underline{F}}_k)$ is dense in $L^1(\underline{\underline{F}}_\infty)=L^1(\underline{\underline{F}})$ (proof ?), and in order to prove that $\underline{\underline{H}}=L^1(\underline{\underline{F}})$ we must only show that $\underline{\underline{H}}$ is closed.

Let X belong to the closure of $\underline{\underline{H}}$. Choose random variables $X^k\in\underline{\underline{H}}$ such that $\|X-X^k\|_1 \leq 2^{-k}$. The maximal theorem (n°5) then gives

(13.2) $$P\{ \sup_n |\overline{X}_n - \overline{X}^k_n| \geq \frac{1}{k} \} \leq k2^{-k} .$$

Since $\sum k2^{-k}<\infty$, the BOREL-CANTELLI lemma implies that the sample function $\overline{X}^k_\cdot(\omega)$, for a.e. ω, converges uniformly to $\overline{X}_\cdot(\omega)$ as $k\to\infty$. Since the property of having a limit at infinity is preserved under uniform convergence, we have proved that $\overline{X}$ converges a.s..

Next we have, the L^1-norm of the conditional expectation operators being 1

(13.3) $$\|X-\overline{X}_n\|_1 \leq \|X-X^k\|_1 + \|X^k-\overline{X}^k_n\|_1 + \|\overline{X}^k_n-\overline{X}_n\|_1 \leq 2^{-k} + \|X^k-X^k_n\|_1 + 2^{-k}$$

which is $\leq 3.2^{-k}$ for n large enough. Thus $\overline{X}_n$ converges to X in L^1, and we have proved that $\underline{\underline{H}}$ is closed. This settles the case p=1, $\underline{\underline{F}}=\underline{\underline{F}}_\infty$. If p=1, $\underline{\underline{F}}\neq\underline{\underline{F}}_\infty$, and $Y\in L^1(\underline{\underline{F}})$ closes (X_n), apply the above results to $(\Omega,\underline{\underline{F}}_\infty,P)$, and the random variable $E[Y|\underline{\underline{F}}_\infty]$ which closes (X_n) on this space. Finally , if Y belongs to L^p, p>1, we have $\sup_n \|X_n\|_p \leq \|Y\|_p$ (the conditional expectation operators are norm decreasing in L^p). Hence $\|\sup_n |X_n|\|_p \leq q\|Y\|_p$ (theorem 8) Therefore dominated convergence takes place in L^p, and X_n converges a.s. and in L^p. The last sentence of the theorem is obvious : if Y is $\underline{\underline{F}}_\infty$-measurable, then $X_\infty = E[Y|\underline{\underline{F}}_\infty]=Y$.

14. REMARK. Let B be a BANACH space, with norm $|\ |$. The preceding proof extends very nicely to the theory of B-valued martingales. For simplicity, we shall consider only the case p=1, leaving the case $1<p<\infty$ to the reader. We first recall some facts on integration

a) If $f : \Omega \to B$ is measurable (B is given its Borel σ-field) then $|f| : \Omega \to \mathbb{R}_+$ is measurable, and we set $\| f \|_1 = \int |f(\omega)| P(d\omega)$

b) If B isn't assumed to be separable, one shouldn't define $L^1(B)$ as the space of all measurable f such that $\|f\|_1 < \infty$, but rather as follows : Let $\underline{\underline{E}}$ be the space of all <u>elementary random variables</u>

$$f(\omega) = \sum_0^m f_n(\omega).a_n \quad (m \in \underset{\sim}{N} ,\ f_n \in L^1(\mathbb{R}) ,\ a_n \in B)$$

Then g measurable belongs to $L^1(B)$ iff there exists a sequence (g_k) of elementary r.v. such that $\|g-g_k\|_1 \to 0$. $L^1(B)$ is a BANACH space.

c) Conditional expectation operators $E[.|\underline{\underline{F}}_n]$ are defined on $\underline{\underline{E}}$ in the obvious way , and extended to $L^1(B)$ by continuity. The definitions of martingales, closed martingales... extend trivially.

Look now at the above proof. We first assume that $\underline{\underline{F}} = \underline{\underline{F}}_\infty$, and define $\underline{\underline{H}}$ without change, except that it consists of B-valued integrable random variables. $\underline{\underline{H}}$ contains $\underline{\underline{E}}$, from the real case, thus $\underline{\underline{H}}$ is dense. The proof that $\underline{\underline{H}}$ is closed requires no change, since the process $|\overline{X}_n - \overline{X}_n^k|$ is a <u>positive submartingale</u>, to which the maximal theorem 5 applies. And the end of the proof again is valid without modification. Thus theorem 13 is valid for B-valued martingales. This is a simple and useful result, and should be known even to people that do not care for vector valued things !

The next theorem is a criterion for closability, which is false for general BANACH space valued martingales.

15. THEOREM. <u>A martingale</u> X <u>is closable in</u> L^1 <u>if and only if it is uniformly integrable</u> , <u>closable in</u> L^p $(1<p<\infty)$ <u>if and only if is bounded in</u> L^p .

PROOF. If $X=(X_n)$ is closable in L^1, then it converges in L^1 (n°13) and hence is uniformly integrable ([1], chap.II, n°21). Conversely, if (X_n) is uniformly integrable there exists a sequence $n_k \uparrow \infty$ such that X_{n_k} converges to some random variable $Y \in L^1$ in the weak* topology of L^1 ([1], chapter II, n°23) - otherwise stated, $\int_A X_{n_k} P$ converges to $\int_A YP$ for every $A \in \underline{\underline{F}}$. Take $A \in \underline{\underline{F}}_m$: then for $n_k > m$ we just have $\int_A X_{n_k} P = \int_A X_m P$, and it follows that $X_m = E[Y|\underline{\underline{F}}_m]$. This means (X_n) is closed by Y.

Any martingale that is closed in L^p is bounded in L^p. Conversely , if (X_n) is bounded in L^p $(1<p<\infty)$, then it is uniformly integrable ([1], chapter II, n°22) , hence closed in L^1, and thus converges a.s. to a closing r.v. X_∞ . FATOU's lemma then implies $\|X_\infty\|_p \leqq \liminf_n \|X_n\|_p < \infty$, and the martingale is closed in L^p.

Our next step consists in proving a.s. convergence for L^1-bounded martingales - a result deeper than theorem 13. Note that convergence no longer takes places in L^1 in general. On our way, we need and prove a very useful lemma, called the KRICKEBERG decomposition lemma :

16. THEOREM. A martingale X is bounded in L^1 if and only if it can be written as a difference of two positive martingales. These martingales $\overset{+}{X}$ and $\overline{X}$ then can be so chosen, that

(16.1) $$\| X \|_1 = \| \overset{+}{X} \|_1 + \| \overline{X} \|_1$$

and such a decomposition then is unique.

PROOF. The "if" part of the theorem is trivial, since every positive martingale is L^1-bounded. To prove the " only if" part, set

$$\overset{+}{X}{}^m_n = E[X^+_{n+m}|\underline{\underline{F}}_n] \qquad , \overline{X}{}^m_n = E[X^-_{n+m}|\underline{\underline{F}}_n] \quad .$$

We have $X_n = \overset{+}{X}{}^m_n - \overline{X}{}^m_n$ for all m ; $\overset{+}{X}{}^{m+1}_n = E[X^+_{n+m+1}|\underline{\underline{F}}_n] = E[X^+_{n+m+1}|\underline{\underline{F}}_{n+m}|\underline{\underline{F}}_n] \geqq [X^+_{n+m}|\underline{\underline{F}}_n] = \overset{+}{X}{}^m_n$. Therefore $\overset{+}{X}{}^m_n$ increases with m and, its expectation being bounded by $\|X\|_1$, converges to an integrable random variable $\overset{+}{X}_n$. Similarily $\overline{X}{}^m_n$ converges to a positive, integrable random variable $\overline{X}_n$, and we have $X_n = \overset{+}{X}_n - \overline{X}_n$. We also have

$$E[\overset{+}{X}{}^{m-1}_{n+1}|\underline{\underline{F}}_n] = E[X^+_{m+n}|\underline{\underline{F}}_{n+1}|\underline{\underline{F}}_n] = \overset{+}{X}{}^m_n \quad .$$

Letting $m \to \infty$ we find that $\overset{+}{X}$ is a martingale. So is $\overline{X}$, and (16.1) is very easy to check. If we have anoter representation X=Y-Z of X as a difference of positive martingales, then $Y_{n+m} \geq X^+_{n+m}$ for all n,m , from which we deduce $Y_n \geqq \overset{+}{X}{}^m_n$ and $Y_n \geqq \overset{+}{X}_n$. Similarily, $Z \geqq \overline{X}$.
The uniqueness statement follows at once.

17. THEOREM. Any L^1-bounded martingale converges a.s. to an a.s. finite random variable.

PROOF. According to 16, it suffices to prove theorem 17 for a positive martingale X . Let then N be an integer, set

$$T = \inf \{ m : X_m \geq N \}$$

and denote by Y the martingale X stopped at T. The random variable X_T is defined only on $\{T<\infty\}$; we set $X_T=0$ on $\{T=\infty\}$, and remark that X_T is integrable, since $E[X_T] \leq \liminf_n E[X_{T\wedge n}] = E[X_0]$ (n°3). Y is dominated by the integrable random variable $N \vee X_T$, since $Y_n \leq N$ for $n<T$, $Y_n=X_T$ for $n \geq T$. Therefore Y converges a.s. to an integrable r.v., by theorems 15 and 13. On the other hand, $Y_.(\omega)=X_.(\omega)$ if $T(\omega) =+\infty$, and thus $X_.$ also converges a.s. on this set, whose probability is arbitrarily close to 1 if N is large enough (n°5).

Note that if X is L^1-bounded, $\lim_n X_n$ is not only finite, but also integrable (FATOU's lemma).

We conclude this sequence of theorems with the supermartingale case. We remark that, if $X=(X_n)$ is a supermartingale, then $(X_n^-) = (-(X_n \wedge 0))$ is a submartingale, and therefore $E[X_n^-]$ increases with n.

18. THEOREM. Let X be a supermartingale such that $\sup_n E[X_n^-]<\infty$. Then X converges a.s. to a finite (even integrable) random variable.

PROOF. Substracting X_0 if necessary, we assume that $X_0=0$. Since $E[X_0] \geq E[X_n] \geq -E[X_n^-]$, $E[X_n]$ remains bounded, and so does $E[|X_n|] = E[X_n]+2E[X_n^-]$. Consider DOOB's decomposition $X=M-A$, where A is increasing with $A_0=0$, $M=X+A$ is a martingale. We have $M_n^- \leq X_n^-$, therefore $E[M_n^-]$ remains bounded. Since $E[M_n]=E[M_0]=0$, $E[|M_n|]$ is $2E[M_n^-]$, and M is L^1-bounded. According to theorem 17, M converges a.s. to an integrable random variable. On the other hand, A_n increases, and $E[A_n]=E[M_n]-E[X_n]$ remains bounded, therefore A_n converges a.s. to an integrable random variable. The theorem follows.

COMMENT. Theorem 13 for BANACH spaces is due to CHATTERJI [12]. The easier proof given here was, according to CHATTERJI [13], first published by BILLINGSLEY. It should be pointed out that the whole theory of convergence extends to martingales with values in reflexive BANACH spaces. The extension of theorem 15 to that case is due to SCALORA [14], that of theorem 17 (which is deeper) to IONESCU-TULCEA [15] (but a simpler proof is given in CHATTERJI [13]). These references are given to help the interested reader, but we will not talk any more about BANACH spaces : our concern is with the real case, and we generalized th.13 only because it was so simple.

CONVERGENCE AND UPCROSSINGS

In spite of the possibility of proving convergence without using the upcrossing inequality (inequality due to DOOB for martingales, to SNELL [16] for supermartingales), upcrossings are one of the basic tools of martingale theory, and must be considered with respect : they can yield much more than convergence, particularly in the continuous case. We are going to use the nice proof given by DUBINS in [17].

19. Let $X=(X_n)$ be a real valued process on Ω. Let a and b be two numbers such that $a<b$. Define the following random variables :

(19.1)
$$\begin{aligned} T_0 &= \inf \{ n : X_n<a \} \\ T_1 &= \inf \{ n : n>T_0 , X_n>b \} \\ T_2 &= \inf \{ n : n>T_1 , X_n<a \} \\ T_3 &= \inf \{ n : n>T_2 , X_n>b \} \\ &\dots\dots \end{aligned}$$

The random variables $T_0, T_2, \dots T_{2k}$ have no particular names, but $T_1, T_3, \dots T_{2k-1} \dots$ are called the first, second,... k-th time an upcrossing of $[a,b]$ is completed (by the sample function $X_{\bullet}$). The total number of completed upcrossings , that is of integers k such that $T_{2k-1}<\infty$, is a random variable, which can be denoted by U_a^b . More explicit notations are useful from time to time, to indicate the dependence on the process, for instance $U_a^b(X ; \omega)$. We shall not try to be systematic about it.

Related definitions are : upcrossings of $]a,b[$ (strict inequalities $X_n<a$, $X_n>b$... in (19.1) are replaced by loose ones, $X_n\leqq a$, $X_n\geqq b$ ...) , downcrossings of $[a,b]$ (or $]a,b[$). We leave their definition to the reader's imagination, pointing out as a check that

(19.2)
$$D_a^b(X ; \omega) = U_{-b}^{-a}(-X ; \omega)$$

We now prove the main inequality :

20. THEOREM. Let $X=(X_n)$ be a supermartingale stopped at time N. Then

$$(20.1)\qquad P\{U_a^b>p\} \leqq \frac{1}{b-a}\int_{\{U_a^b=p\}}(X_N-a)^-\,P \qquad (\,p\in\mathbb{N}\,)$$

If we sum over p, we get the upcrossing inequality in DOOB's form :

21. COROLLARY . $E[U_a^b] \leqq \frac{1}{b-a}E[(X_N-a)^-]$.

PROOF. We can assume that a=0. We define two stopping times

S : the first time n that $X_n<0$ after the p-th upcrossing was completed,

T : the time the (p+1)-th upcrossing is completed.

We also set $T'=T\wedge N$. The inequalities

$$(20.2)\qquad bP\{\,U_0^b>p\,\} \leqq \int_{\{U_0^b>p\}} X_T \leqq \int_{\{S<\infty\}} X_{T'}^+$$

follow easily from the definition of upcrossings. Next we apply the supermartingale inequality at the bounded stopping times $S'=S\wedge N$, $T'=T\wedge N$: $\{S<\infty\}$ belongs to $\underline{\underline{F}}_S$ and to $\underline{\underline{F}}_N$ (since the process is stopped at N), hence to $\underline{\underline{F}}_{S'}$, and on this set $X_{S'}=X_S\leqq 0$. Consequently

$$(20.3)\qquad 0 \geqq \int_{\{S<\infty\}} X_{S'} \geqq \int_{\{S<\infty\}} X_{T'}$$

We deduce from it the following estimate for the right side of (20.2)

$$(20.4)\qquad \int_{\{S<\infty\}} X_{T'}^+ \leqq \int_{\{S<\infty\}} X_{T'}^- = \int_{\{S<\infty\,,X_{T'}<0\}} X_{T'}^- \leqq \int_{\{S<\infty\,,X_{T'}<0\}} X_{T'}^-$$

since (X_n^-) is a submartingale . But if the (p+1)-th upcrossing is completed we must have $X_{T'}\geqq b$, therefore on $\{S<\infty\,,\ X_{T'}<0\}$ the (p+1)-th upcrossing is never completed, and U_0^b is exactly p. Finally

$$(20.5)\qquad \int_{\{S<\infty\,,X_{T'}<0\}} X_N^- \leqq \int_{\{U_0^b=p\}} X_N^-$$

Putting (20.2), (20.4) and (20.5) in one single chain completes the proof.

22. REMARKS. 1) As in the case of the maximal lemma (n°6), inequality (20.1) can be conditioned

$$(22.1)\qquad P\{U_a^b>p\,|\,\underline{\underline{F}}_0\} \leqq \frac{1}{b-a}\,E[(X_N-a)^-I_{\{U_a^b=p\}}\,|\,\underline{\underline{F}}_0\,] \quad \text{a.s.}$$

Also a,b,p can be replaced by $\underline{\underline{F}}_0$-measurable functions satisfying some obvious conditions.

2) Assume X,a are positive. Then $(X_N-a)^- \leqq a$, and (20.1) gives

$$P\{U>p\} \leqq \frac{a}{b-a}P\{U=p\}$$

(where we write U instead of U_a^b for simplicity). This can be written as

$$P\{U>p\} \leqq \frac{a}{b}(P\{U>p\}+P\{U=p\}) = \frac{a}{b}P\{U>p-1\}$$

and finally $P\{U>p\} \leqq (\frac{a}{b})^p P\{U>0\}$. This last probability can be computed directly , as in the proof of the maximal lemma, getting the estimate

$$P\{U>0\} \leqq \frac{1}{b}E[X_0 \wedge a]$$

which is sharper in this case than (20.1). We get finally

(22.2) $$P\{U_a^b>p\} \leqq \frac{E[X_0\wedge a]}{b} \left(\frac{a}{b} \right)^p$$ for a positive supermartingale

an inequality which is also due to DUBINS [18]. Note the integrated formula

(22.3) $$E[U_a^b] \leqq \frac{E[X_0\wedge a]}{b-a} \leqq \frac{a}{b-a}$$

The right side doesn't depend on X at all (provided X is positive) ! This rather amazing fact was first discovered by HUNT.

23. To illustrate the power of the upcrossing inequality, we prove the a.s. convergence theorem 18. Let X be a supermartingale such that $\sup_n E[X_n^-] < \infty$, and let a,b be two numbers such that a<b. We have according to theorem 21, stopping X at time N and then letting $N\to\infty$.

$$E[U_a^b] \leqq \frac{1}{b-a} \sup_N E[(X_N-a)^-] \leqq \frac{1}{b-a}(a^+ + \sup_N E[X_N^-]) < \infty$$

Therefore U_a^b is a.s. finite. On the other hand, if ω belongs to the set

$$W(a,b) = \{ \liminf_{n\to\infty} X_n < a, \limsup_{n\to\infty} X_n > b \}$$

then $U_a^b(\omega)=+\infty$. Hence W(a,b) is a null set, and so is the union of all sets W(a,b) over all pairs of rationals (a,b) with a<b. This union is just $\{ \limsup_{n\to\infty} X_n > \liminf_{n\to\infty} X_n \}$, the set where $X_.$ doesn't converge. Finiteness (integrability) of the limit is deduced from FATOU's lemma and the inequality $E[|X_n|] = E[X_n] +2E[X_n^-] \leqq E[X_0]+2E[X_n^-]$.

CONVERGENCE, INFINITY TO THE LEFT

We have been considering only processes indexed by $\mathbb{N}$, that is with infinity "to the right". What about processes indexed by $\mathbb{Z}$ or $-\mathbb{N}$, and convergence problems as $n \to -\infty$? The basic result is the following theorem, due to DOOB, thanks to which everything becomes very simple.

24. THEOREM. Let $X=(X_n)_{n\in -\mathbb{N}}$ be a supermartingale relative to an increasing family of σ-fields $(\underline{\underline{F}}_n)_{n\in -\mathbb{N}}$. Then either $\lim_{n\to-\infty} E[X_n] = +\infty$, or X is uniformly integrable.

PROOF. $E[X_n]$ is a decreasing function of n, therefore the limit $A = \lim_n E[X_n]$ exists as $n\to-\infty$. Let us assume A is finite, and prove that X is uniformly integrable. We remark that the martingale $E[X_0|\underline{\underline{F}}_n]$ is uniformly integrable (see lemma 25 below). Substracting it from X , we are reduced to the case where X is positive. Let ε be >0, choose k large enough so that $A-E[X_{-k}] < \varepsilon/2$. The finite family $X_0, X_{-1}, \ldots, X_{-k}$ being uniformly integrable, it is sufficient to show that for c large enough

$$\sup_{n<-k} \int_{\{X_n>c\}} X_n . P \leq \varepsilon$$

Now for $n<-k$ we have from the supermartingale inequality

$$\int_{\{X_n>c\}} X_n = E[X_n] - \int_{\{X_n\leq c\}} X_n \leq E[X_n] - \int_{\{X_n\leq c\}} X_{-k}$$
$$= E[X_n]-E[X_{-k}] + \int_{\{X_n>c\}} X_{-k}$$

$E[X_n]$ lies somewhere between $E[X_{-k}]$ and A, therefore the first difference is $< \varepsilon/2$. On the other hand, $P\{X_n>c\} \leq \frac{A}{c}$ (CHEBYSHEV inequality), and the last integral becomes $\varepsilon/2$ for c large enough. The theorem is proved.

Here is the lemma we have used - it extends also part of theorem 15. We shall use it again later on.

25. LEMMA . Let X be an integrable random variable. Then the family of all random variables $E[X|\underline{\underline{G}}]=X_{\underline{\underline{G}}}$, where $\underline{\underline{G}}$ is a sub-σ-field of $\underline{\underline{F}}$, is uniformly integrable.

PROOF. Replacing X by $|X|$ if necessary, we can assume X is positive. Then we have

$$\int_{\{X_{\underline{\underline{G}}}>c\}} X_{\underline{\underline{G}}} = \int_{\{X_{\underline{\underline{G}}}>c\}} X$$

$$P\{X_{\underline{\underline{G}}}>c\} \leq \frac{1}{c}E[X_{\underline{\underline{G}}}] = \frac{1}{c}[X]$$

Hence the first integral is $<\varepsilon$ for c large enough.

The convergence theorem is very easy :

26. THEOREM. Let $X=(X_n)_{n\in -\mathbb{N}}$ be a supermartingale, such that $\lim_{n\to -\infty} E[X_n] < +\infty$. Then X_n converges a.s. and in L^1 to a random variable $X_{-\infty}$. In the martingale case, $X_{-\infty}=E[X_0|\bigcap_n \underline{\underline{F}}_n]$.

PROOF. To prove a.s. convergence, one uses the upcrossing inequality $E[U_a^b] \leq \frac{1}{b-a}E[(X_0-a)^-]$
between times $-N$ and 0, and lets $N\to+\infty$; then one proceeds as in n°23. Since $X_n \to X_{-\infty}$ a.s., and the sequence is uniformly integrable, convergence takes place in L^1 ([1], chapter II, n°21). The last sentence is obvious. **Applying these results to $X\wedge m$ and letting $m\to\infty$ will show that a.s. convergence (to a possibly infinite r.v.) holds even if $\lim E[X_n]=+\infty$.**

27. REMARKS. a) If $X_0 \in L^p$, theorem 8 implies that $X_n \to X_{-\infty}$ in L^p.

b) There exists also a " proof without upcrossings" for this theorem. We sketch it for the martingale case : the space $\underline{\underline{H}}$ of all X such that $E[X|\underline{\underline{F}}_n]$ converges a.s. to $E[X|\bigcap_n \underline{\underline{F}}_n]$ is closed in L^1 (maximal lemma) and dense in L^1 since[1] it contains all X which are $\bigcap_n \underline{\underline{F}}_n$-measurable, and all $X=Y-E[Y|\underline{\underline{F}}_n]$ ($Y\in L^1$). Hence $\underline{\underline{H}}=L^1$.

c) In the martingale case, theorem 26 is true for processes taking values in a BANACH space B. Define $\underline{\underline{H}}$ as in b), X being an element of $L^1(B)$. Then $\underline{\underline{H}}$ is closed in $L^1(B)$, and dense since it contains (from the real case) all elementary random variables (see n°14).

OPTIONAL SAMPLING, UNBOUNDED STOPPING TIMES

28. We return to our usual index set $\mathbb{N}$. Let X be a supermartingale, and let S,T be two stopping times such that $S\leq T$. The basic theorem 3 asserts that, whenever T (and S !) is bounded, X_S and X_T are integrable and

(28.1) $\qquad X_S \geq E[X_T|\underline{\underline{F}}_S]$ a.s.

We would like to extend this supermartingale inequality to more general stopping times (of course, if we want to consider also stopping times which assume the value $+\infty$, we must properly define

[1] A small L^2-argument works, $L^1\cap L^2$ being dense in L^1.

the value X_∞ of the process at infinity, and the σ-field $\underline{\underline{F}}_\infty$ - otherwise $\underline{\underline{F}}_S$ doesn't make sense). The general idea is that no " best " result of this kind can exist : suitable restrictions on X imply (28.1) for suitable classes of stopping times, no restriction at all corresponding to bounded stopping times , and uniform integrability implying (28.1) for all stopping times. We shall not explore systematically this field : we give the main theorem, and a small sample of the more special results. We need the following definition :

DEFINITION. The supermartingale X is closed (to the right) by a random variable $X_\infty \in L^1$ if, for all n ,

(28.2) $$X_n \geqq E[X_\infty | \underline{\underline{F}}_n] \quad \text{a.s.}$$

In many cases, X_∞ may be defined as $\lim_n X_n$, but other choices may be convenient. For instance , $\lim_n X_n$ is a possible choice for X_∞ if X is uniformly integrable (L^1-convergence), and also if X is positive (FATOU's lemma), but in this latter case 0 also closes X, and so does any random variable orthogonal to $\bigvee_n \underline{\underline{F}}_n$.

The comparison between definitions 12 and 28 should be done : a martingale is also a supermartingale. It may, however, be closable as a supermartingale and not as a martingale, since any closed martingale is uniformly integrable (n°15).

29. THEOREM. Let $X=(X_n)_{n\in\underline{\underline{N}}\cup\{+\infty\}}$ be a closed supermartingale, and let S and T be (possibly infinite) stopping times such that $S\leqq T$. Then X_S and X_T are integrable and $X_S \geqq E[X_T|\underline{\underline{F}}_S]$ a.s.. If X is a martingale, closed by X_∞ as a martingale, then $X_S=E[X_T|\underline{\underline{F}}_S]=E[X_\infty|\underline{\underline{F}}_S]$ a.s.[1]

PROOF. Set $Y_n=E[X_\infty|\underline{\underline{F}}_n]$, $Z_n=X_n-Y_n$, $Y_\infty=X_\infty$, $Z_\infty=0$. Then $(Y_n)_{n\in\underline{\underline{N}}\cup\{+\infty\}}$ is a uniformly integrable martingale, closed as a martingale, and $(Z_n)_{n\in\underline{\underline{N}}\cup\{+\infty\}}$ is a positive supermartingale closed by 0. Since X=Y+Z, it will be sufficient to prove the theorem for Y and Z separately. The study for Y will also give us the last sentence of the above statement.

a) Set $S_n=S$ on $\{S\leqq n\}$, $S_n=+\infty$ on $\{S>n\}$. The set $\{0,1,\dots,n,+\infty\}$ is order isomorphic to $\{0,1,\dots,n,n+1\}$, and therefore theorem 3 gives us

$$Y_{S_n} = E[Y_\infty | \underline{\underline{F}}_{S_n}] \quad \text{a.s.}$$

[1] We may define $\underline{\underline{F}}_\infty=\bigvee_n \underline{\underline{F}}_n$

The sets $\{S\leqq n\}$, $\{S=\infty\}$ belong to $\underline{\underline{F}}_S$, hence to $\underline{\underline{F}}_{S_n}$, and on these sets the functions Y_{S_n} and Y_S, the operators $E[.|\underline{\underline{F}}_{S_n}]$ and $E[.|\underline{\underline{F}}_S]$ agree. Therefore $Y_S=E[Y_\infty|\underline{\underline{F}}_{S_n}]$ a.s. on $\{S\leqq n$ or $S=\infty\}$. Letting $n\to\infty$ we get $Y_S=E[Y_\infty|\underline{\underline{F}}_S]$ a.s., which implies in particular the integrability of Y_S. Finally

$$Y_S=E[Y_\infty|\underline{\underline{F}}_S] = E[Y_\infty|\underline{\underline{F}}_T|\underline{\underline{F}}_S] = E[Y_T|\underline{\underline{F}}_S] \text{ a.s..}$$

b) Z_S is integrable according to FATOU's lemma applied to the inequality $E[Z_0] \geqq E[Z_{S_n}]$. Define T_n in the same way as S_n. We have from theorem 3, as in part a)

$$Z_{S_n} \geqq E[Z_{T_n}|\underline{\underline{F}}_{S_n}] \text{ a.s.}$$

and therefore, as above

$$Z_S \geqq E[Z_{T_n}|\underline{\underline{F}}_S] \text{ a.s. on } \{S\leqq n \text{ or } S=\infty\}$$

letting $n\to\infty$ and remarking that $Z_{T_n}\uparrow Z_T$ gives the result.

The following trivial corollary is worth mentioning : in our future language, it can be stated <u>every uniformly integrable martingale belongs to the class</u> (D).

30. THEOREM. <u>Let</u> $X_n=E[X_\infty|\underline{\underline{F}}_n]$ <u>be an uniformly integrable martingale</u>. <u>Then all random variables</u> X_T (<u>for all stopping times</u> T) <u>are uniformly integrable</u> .

PROOF. $X_T=E[X_\infty|\underline{\underline{F}}_T]$, then apply lemma 25.

Another amusing consequence (pointed out to me by J.LAZARO). Conditioning X with respect to $\underline{\underline{F}}_T$ amounts to stopping at time T the martingale $E[X|\underline{\underline{F}}_n]$. If S and T are two stopping times, conditioning w.r. to $\underline{\underline{F}}_T$ and then to $\underline{\underline{F}}_S$ amounts to stopping at T and then at S, that is, to stopping at $T\wedge S$. Thus the conditional expectation operators $E[.|\underline{\underline{F}}_T]$ and $E[.|\underline{\underline{F}}_S]$ commute.

31. The following theorem is given just as a sample, and can be omitted.

The idea is as follows : X is a supermartingale, and is no longer assumed to be closed. Hence our two stopping times $S\leqq T$ are supposed to be a.s. finite. Instead of the definitions of S_n and T_n in n°29, set $S_n=S\wedge n$, $T_n=T\wedge n$. Let A belong to $\underline{\underline{F}}_S$; then $A\cap\{S\leqq n\} = A\cap\{S\leqq S_n\}$ belongs to $\underline{\underline{F}}_{S_n}$, and theorem 3 gives us

(31.1) $$\int_{A\cap\{S\leqq n\}} X_{S_n}.P \geqq \int_{A\cap\{S\leqq n\}} X_{T_n}.P$$

We let now $n\to\infty$, <u>and the only thing we must prove is L^1 convergence of X_{S_n}, X_{T_n} to X_S, X_T</u>.

Set $S_n=H$, $S=K$, and define $V_n=I_{\{H<n\leq K\}}$; this is a predictable process, and

(31.1) $E[|X_K-X_H|]=E[|\sum_n V_n(X_n-X_{n-1})|] \leq E[\sum_n V_n|X_n-X_{n-1}|]$

$$= E[\sum_n V_n E[|X_n-X_{n-1}||\underline{\underline{F}}_{n-1}]]$$

Making a strong assumption, we get the following result

THEOREM. <u>Assume that for every</u> n $E[|X_n-X_{n-1}||\underline{\underline{F}}_{n-1}]\leq M$. <u>Then the optional sampling theorem holds for integrable stopping times.</u>

PROOF. The last expectation above is smaller than $ME[\sum_n V_n]=$ $ME[K-H]$. If $S=K$ is integrable, $S_n\to S$ in L^1 and $X_{S_n}\to S$ in L^1 from (31.1).

Integrability of S and T will imply the validity of the optional sampling theorem under other related hypotheses. For shortness set $E[|X_n-X_{n-1}||\underline{\underline{F}}_{n-1}]=\xi_n$. Instead of assuming that the sequence $(\xi_n)=\xi$ is uniformly bounded, assume it belongs to $L^p(\ell^p)$ for some p, $1<p<\infty$, that is $E[\sum_n \xi_n^p]<\infty$, and denote by q the conjugate exponent as usual. Applying twice HÖLDER's inequality, we get first

$$\sum V_n\xi_n \leq (\sum V_n^q)^{1/q}(\sum \xi_n^p)^{1/p}$$

then

$$E[|X_K-X_H|]\leq E[\sum V_n\xi_n]\leq (E[\sum V_n^q])^{1/q}(E[\sum \xi_n^p])^{1/p}$$
$$=(E[K-H])^{1/q}(E[\sum \xi_n^p])^{1/p}=C(E[K-H])^{1/q}$$

and the conclusion follows as above.

<u>THE RIESZ DECOMPOSITIONS</u>

There is a far reaching analogy between <u>martingales</u> and <u>harmonic functions</u>, <u>super</u>martingales and <u>super</u>harmonic functions. In classical potential theory (of the unit disc, for instance), GREEN potentials of positive measures can be characterized as positive superharmonic functions p, such that the only positive harmonic function $h\leq p$ is 0, and every positive superharmonic function can be decomposed uniquely into a sum of a positive harmonic function and a potential. This is the classical F.RIESZ decomposition. Note that its apparently unambiguous phrasing is slightly misleading :

the intuitive meaning of potentials is that of being superharmonic functions which " are zero at the boundary" . Then the POISSON kernel at a boundary point x, a harmonic function which is 0 at the boundary except at x, may be considered as a potential, while on the other hand the GREEN kernel at an inner point x, a potential, becomes harmonic if we ignore its pole, a polar point. The same ambiguity will appear in the decompositions below.

We begin with the most obvious one.

32. THEOREM. *Let X be a positive supermartingale. Set $X_\infty = \lim_n X_n$ and decompose it into*

$$(31.1) \qquad X_n = (X_n - E[X_\infty | \underline{\underline{F}}_n]) + E[X_\infty | \underline{\underline{F}}_n]$$

This is the only decomposition X=Y+Z, where Z is an uniformly integrable martingale, and Y a positive supermartingale such that $\lim_n Y_n = 0$ a.s..

PROOF. We have $X_n \geq E[X_\infty | \underline{\underline{F}}_n]$ from FATOU's lemma, therefore Y is a positive martingale. Since both X_n and $E[X_\infty | \underline{\underline{F}}_n]$ converge to X_∞ (n°13), their difference Y_n tends to 0. On the other hand, Z is uniformly integrable (n°15), and therefore the decomposition possesses the above stated properties. If X=Y+Z=Y'+Z' are two such decompositions, Z-Z' is an uniformly integrable martingale which converges a.s. to 0, and therefore is 0 (n°15). Thus Z=Z', Y=Y' and uniqueness is proved.

REMARK. One can of course write (32.1) whenever X_∞ exists a.s. and is integrable, but there is no reason why $X_n - E[X_\infty | \underline{\underline{F}}_n]$ should be positive in this case.

The true RIESZ decomposition comes next

33. THEOREM. *Let X be a supermartingale such that $\lim_{n \to \infty} E[X_n] > -\infty$. Then X can be represented uniquely as a sum X=Y+Z, where Z is a martingale , and Y is a positive supermartingale such that $Y_n \to 0$ in L^1. Z is positive if X is positive.*

PROOF. Consider the martingales $Z_n^m = E[X_m | \underline{\underline{F}}_n]$. It follows at once from the supermartingale property of X that, once m becomes larger than n, Z_n^m decreases as m increases. On the other hand, $E[Z_n^m] = E[X_m]$ is bounded from below, and therefore Z_n^m converges in L^1 to a random variable Z_n as $m \to \infty$. It is obvious that $Z_n \geq 0$ if X is positive. LEBESGUE's theorem implies that $Z=(Z_n)$ is a martingale, and therefore Y=X-Z is a supermartingale. For $m \geqq n$ we have $Z_n^m \leqq X_n$, so that $Y \geqq 0$. We have

$$E[Y_n] = \lim_{m\to\infty} E[X_n - Z_n^m] = \lim_{m\to\infty} E[X_n - X_m] = E[X_n] - \lim_{m\to\infty} E[X_m]$$

which tends to 0 as $n\to\infty$. Thus the decomposition X=Y+Z possesses the above properties. To prove uniqueness, consider two such decompositions X=Y+Z=Y'+Z'. Then Z-Z' is a martingale which converges to 0 in L^1, and therefore is 0. Thus Z=Z' and Y=Y'.

SECOND PROOF. Consider DOOB's decomposition X=M-A, where M is a martingale, A an increasing process with $A_0=0$. We have $E[X_n] = E[M_n]-E[A_n] = E[X_0]-E[A_n]$, and the hypothesis amounts to the fact that $E[A_n]$ remains bounded. Therefore $A_\infty = \lim_n A_n$ belongs to L^1. The decomposition X=Y+Z

$$Y_n = E[A_\infty | \underline{\underline{F}}_n] - A_n \quad , \quad Z_n = M_n - E[A_\infty | \underline{\underline{F}}_n]$$

possesses the desired properties, and therefore, according to uniqueness, is the same decomposition as above.

In the continuous case, these two methods of proof will lead to _different_ decompositions of X.

In analogy with the definition of a potential in analysis, we set

34.DEFINITION. _A_ potential _is a positive supermartingale_ X _which doesn't dominate any positive martingale except_ 0.

35. THEOREM. _A positive supermartingale_ X _is a potential if and only if_ $\lim_{n\to\infty} E[X_n]=0$.

PROOF. If X dominates a positive martingale $Z\neq 0$, $E[X_n]$ dominates the constant $E[Z_n]>0$. Conversely, if $\lim_n E[X_n]>0$, the martingale Z in the RIESZ decomposition of X, which is dominated by X, cannot be equal to 0.

Note that if X is a potential, the limit X_∞ exists and is a.s. equal to 0, but the converse isn't true : this is precisely the difference between the decompositions 32 and 33.

A CURIOUS RESULT

We end this section with a remark of B. DAVIS [19], which doesn't seem to have any particular use, but deserves not to be forgotten. The proof below isn't the original one, which was an application of the maximal lemma : it is much simpler, and due to J.L.DOOB.

36. THEOREM. Let X be a positive supermartingale, and set

$$X^* = \sup_n X_n \ , \quad X_k^* = \sup_{n \leq k} X_n$$

Let Φ be a random variable, $\underline{\underline{F}}_k$-measurable and a.s. dominated by X^* . Then $\Phi \leq X_k^*$ a.s..

PROOF. Consider the supermartingale $Y_n = X_{k+n}$ $(n \geq 0)$ relative to the family of σ-fields $\underline{\underline{G}}_n = \underline{\underline{F}}_{k+n}$, and the stopping time $T = \inf\{ n : Y_n \geq \Phi - \varepsilon \}$. Then T is a.s. finite on the set $\{X_k^* < \Phi\}$, (Y_0, Y_T) is a supermartingale and we have

$$\int_{\{X_k^* < \Phi\}} Y_T \leq \int_{\{X_k^* < \Phi\}} Y_0 \leq \int_{\{X_k^* < \Phi\}} \Phi \quad (\text{ since } Y_0 = X_k \leq X_k^*)$$
$$\leq \int_{\{X_k^* < \Phi\}} (Y_T + \varepsilon)$$

The difference between the second and third integrals hence is <ε. Since ε is arbitrary , $\int_{\{X_k^* < \Phi\}} (\Phi - Y_0) = 0$. On the other hand, $Y_0 = X_k \leq X_k^*$, and this equality is absurd unless $P\{X_k^* < \Phi\} = 0$.

Positivity of X hasn't really been used : the important thing is the validity of the stopping theorem for arbitrary stopping times .

§ 2 . A SAMPLE OF MODERN MARTINGALE THEORY

This section contains miscellaneous results , the best known of which are the BURKHOLDER inequalities. We haven't tried to cover all modern developments. Our guide has been, of course, the relations to the theory of stochastic integrals.

CLASS (D) PROCESSES

This rather important notion was introduced by DOOB in potential theory (as a characterization of DIRICHLET solutions).

37. DEFINITION. A process X belongs to the class (D) if all random variables X_T , T ranging over the set of all finite stopping times, are uniformly integrable.

Finite may be replaced by bounded in this definition, since $X_T = \lim_n X_{T\wedge n}$ if T is finite. Any process such that $X^* = \sup_n |X_n|$ belongs to L^1 belongs to the class (D). A less trivial example was given in n° 30 : uniformly integrable martingales belong to the class (D). This result is generalized in the following theorem - misleadingly so, since the statement will be wrong for continuous time supermartingales.

38. THEOREM. Any uniformly integrable supermartingale X belongs to the class (D).

PROOF. Consider DOOB's decomposition , X=M-A . We have $E[A_\infty] = \lim_n E[A_n] = \lim_n (E[M_n]-E[X_n]) < \infty$. Therefore A_∞ belongs to L^1, and all random variables A_T , dominated by A_∞ , are uniformly integrable. On the other hand, M=X+A is a uniformly integrable martingale, and therefore (n°30) all M_T are uniformly integrable. The same then is true for all $X_T=M_T+A_T$.

There is in fact a close relation between class (D) processes and martingale theory : if X belongs to the class (D), there is a supermartingale Y of the class (D) above X (and a submartingale of the class (D) below it). Using DOOB's decomposition, it is easy to find a martingale Z of the class (D) such that $|X| \leqq Z$. We aren't going to prove here these results, which are related to optimal stopping theory . The ideas for the construction of Y are those of SNELL [16]. Details can be found in MERTENS [20] and [21].

As a by-product of the preceding proof, we get

39. THEOREM. If X is a potential, the DOOB decomposition X=M-A can be written as

(39.1) $X_n = E[A_\infty | \underline{\underline{F}}_n] - A_n$ ($M_n = E[A_\infty | \underline{\underline{F}}_n]$)

PROOF. Every potential converges to 0 in L^1, and therefore is uniformly integrable. M belongs to the class (D), and $M_\infty = X_\infty + A_\infty = A_\infty$, hence $M_n = E[A_\infty | \underline{\underline{F}}_n]$.

We may set the following definition

40. DEFINITION. An increasing process A (not necessarily predictable or adapted) is said to be integrable if A_∞ belongs to L^1. The process (class (D) potential)

(40.1) $X_n = E[A_\infty - A_n | \underline{\underline{F}}_n]$

is called the potential generated by A.

We leave to the reader the easy proof that X indeed is a positive supermartingale, and converges to 0 in L^1.

LOCAL SUPERMARTINGALES AND MARTINGALES

These notions will be extremely important in later chapters. Here too the discrete situation looks misleadingly simple. The definition of local martingales was given by ITO-S.WATANABE [22] in the continuous time case - they haven't been much considered in discrete time.

41. DEFINITION. An adapted process X is a local martingale (supermartingale) if there exists a sequence T_k of stopping times, increasing a.s. to $+\infty$, and such that for every k the process $(X_{n \wedge T_k} I_{\{T_k > 0\}})$ is a martingale (supermartingale).

Replacing T_k by $T_k \wedge k$, we may even assume that this process is a uniformly integrable martingale (supermartingale). It will be convenient to coin a word for this property : if T is a stopping time and $(X_{n \wedge T} I_{\{T>0\}})$ is a class (D) (= uniformly integrable) martingale or supermartingale, we shall say that T reduces X (understood : to a martingale or supermartingale). The restriction to class (D) processes will be important only in continuous time.

The following theorem is partly false in continuous time . It shows that local martingales are rather trivial extensions of martingales, and at the same time rather interesting . Note the corollary (valid in discrete time only) : every local martingale with integrable random variables is a true martingale.

We leave to the reader the similar result for local supermartingales.

42 . THEOREM. Let X be adapted. Then the following properties are equivalent :

1) X is a local martingale .

2) For every n, $E[|X_{n+1}|\,|\underline{\underline{F}}_n] < \infty$ a.s. and $E[X_{n+1}|\underline{\underline{F}}_n]$, which is meaningful according to the preceding property, is a.s. equal to X_n .

3) X is a martingale transform .

PROOF. Assume 1). Let the stopping times T_k be as in the definition, and A belong to $\underline{\underline{F}}_m$; then so does $A\cap\{T_k>m\}$, and we may apply the martingale property on this set to the process $(X_{T_k\wedge m}\, I_{\{T_k>0\}})$, so that $X_{m+1}I_{\{T_k>m\}}$ and $X_m I_{\{T_k>m\}}$ are integrable and

$$\int_{A\cap\{T_k>m\}} X_{m+1} = \int_{A\cap\{T_k>m\}} X_m$$

Since $\{T_k>m\}$ increases a.s. to Ω as $k\to\infty$, 2) follows at once. Thus 1)=>2).

Assume 2) , set as usual $x_0=X_0$, $x_n=X_n-X_{n-1}$, $W_n=E[|x_n|\,|\underline{\underline{F}}_{n-1}]$ (also $W_0=|x_0|$), and finally

$$V_n = 1/W_n \text{ if } W_n>0 \text{ , } V_n=0 \text{ if } W_n=0$$

This is a predictable process, and we may define the transform $Y=V\cdot X$. Setting $y_0=V_0x_0$, $y_n = Y_n-Y_{n-1}=V_nx_n$, it is very easy to see that

$$E[|y_n|\,|\underline{\underline{F}}_{n-1}] \leqq 1 \text{ , } E[y_n|\underline{\underline{F}}_{n-1}] = 0$$

Thus Y is a true martingale, and $X=W\cdot Y$ is a martingale transform. Otherwise stated, 2)=>3).

Finally, let X be a martingale, V be predictable, and Y be the transform $V\cdot X$. Set $T_k = \inf\{\, n : |V_{n+1}|>k \,\}$, a stopping time according to the predictability of V. Then V stopped at T_k

is bounded by k on $\{T_k>0\}$, so that $(Y_{n\wedge T_k} I_{\{T_k>0\}})$ is a martingale (n°1). Since T_k increases to ∞ with k, Y is a local martingale, 3) implies 1) and the theorem is proved.

Note that any transform of a local martingale still is a local martingale.

Local supermartingales have decompositions like ordinary supermartingales

43. THEOREM. Let X be a local supermartingale. Then X can be uniquely decomposed as a difference M-A, where M is a local martingale, and A is a predictable increasing process with $A_0=0$.

PROOF (sketch) . Take stopping times T_k as in the definition (41), and write the DOOB decomposition of the process $(X_{n\wedge T_k} I_{\{T_k>0\}})$ as M^k-A^k. Then check using uniqueness that these processes can be " pasted together" into single processes M and A. There is also a trivial proof using theorem 42, 2).

44. AN EXAMPLE. Let $(x_n)_{n\geq 1}$ be a sequence of independent random variables with symmetrical distributions, and set $X_0=0$, $X_n=x_1+\dots+x_n$. On some other space, build a wholly independent system of random variables V_n, y_n ($n\geq 0$) with the following laws : the y_n are BERNOULLI random variables ($P\{y_i=+1\}=P\{Y_i=-1\}=\frac{1}{2}$), $V_0=0$, and for $n\geq 1$ V_n has the same law as $|x_n|$. For $n\geq 1$, let $\underline{\underline{F}}_n$ be the σ-field generated by $y_0,\dots,y_n,V_0,\dots,V_n,V_{n+1}$ (for n=0, by y_0,V_0,V_1), so that V is a predictable process, y_{n+1} is independent from V_n, the process $Y_0=0$, $Y_n=y_1+\dots+y_n$ is a martingale, and the processes X and V.Y have the same law, since $V_n y_n$ has the same law as x_n . Thus a sum of independent, symmetrically distributed random variables can be interpreted as a local martingale on a suitable space (if $P\{x_n=0\}=0$ for all n, the construction can be carried out on the original space, setting $V_n=|x_n|$, $y_n=\mathrm{sgn}(x_n)$, $V_0=y_0=0$). The fact that sums of symmetrically distributed independent r.v. are transforms of BERNOULLI processes explains many of their special properties.

This example illustrates also a somewhat paradoxical property of local martingales : in the above construction, set Z= V.Y . Then Z is a local martingale relative to $(\underline{\underline{F}}_n)$, and z_n is independent of $\underline{\underline{F}}_{n-2}$: therefore, if z_n isn't integrable, it shall not be integrable on $\underline{\underline{F}}_{n-2}$ sets of measure 0. Hence the process $(Z_0, Z_2, Z_3, \ldots)$ isn't a local martingale with respect to $(\underline{\underline{F}}_0, \underline{\underline{F}}_2, \underline{\underline{F}}_3, \ldots)$. Similarily, we shall see that if (X_t) is a local martingale in continuous time, then the discrete process (X_n) isn't necessarily a local martingale in discrete time. This is quite different from the situation with true martingales !

ESTIMATING DOOB'S DECOMPOSITION : THE BOUNDED CASE

45. Computations relative to DOOB's decomposition lead to a large number of beautiful inequalities on martingales (which were proved formerly in a much more complicated way). We start here with the simplest case .

Let X be a potential (n°34) . Let us write its DOOB decomposition X=M-A , according to n°39, as

(45.1) $$X_n = E[A_\infty | \underline{\underline{F}}_n] - A_n \quad , \quad M_n = E[A_\infty | \underline{\underline{F}}_n]$$

We then have for any stopping time T , $M_T = E[A_\infty | \underline{\underline{F}}_T]$, and therefore $X_T = E[A_\infty | \underline{\underline{F}}_T] - A_T$ (see n°29). Our purpose is to get estimates of A_∞ from estimates of X. The most elementary inequalities concern bounded potentials : the reader should work out explicitly for himself the case p=2, which is almost trivial and by far the most important one.

THEOREM. Let the potential X be dominated by a constant c. Then we have for any integer[(1)] $p>0$

(45.2) $$E[A_\infty^p] \leq p!c^{p-1}E[X_0] \leq p!c^p$$

and therefore, for $t < \frac{1}{c}$, $\exp[tA_\infty]$ belongs to L^1.

PROOF. We write A_∞ as $\sum_i (A_{i+1} - A_i)$ and raise to the p-th power

$$A_\infty^p = \sum_{i_1, i_2, \ldots, i_p} (A_{i_1+1} - A_{i_1})(A_{i_2+1} - A_{i_2}) \ldots (A_{i_p+1} - A_{i_p})$$

We single out the largest integer among $i_1, \ldots, i_p$, which we call j. Since j can occur in the first,...,p-th position, we have

$$A_\infty^p \leq p \sum_j \sum_{j_1, \ldots, j_{p-1} \leq j} (A_{j_1+1} - A_{j_1}) \ldots (A_{j_{p-1}+1} - A_{j_{p-1}})(A_{j+1} - A_j)$$

We may write this as follows, summing over j first

---

(1) Some other proof should work for real p !

$$p \sum_{j_1,\dots,j_{p-1}} (A_{j_1+1}-A_{j_1})\dots(A_{j_{p-1}+1}-A_{j_{p-1}})(A_\infty - A_{j_1\vee\dots\vee j_{p-1}})$$

We take expectations. All terms except the last one are measurable with respect to $\underline{\underline{F}}_{j_1\vee..\vee j_{p-1}}$. We therefore can replace the last term by its conditional expectation relative to this σ-field, which is $X_{j_1\vee..\vee j_{p-1}} \leqq c$. Thus we get the induction inequality

$$E[A^p_\infty] \leqq cpE[A^{p-1}_\infty]$$

Since $E[A_\infty]=E[X_0]$, (45.2) follows at once.

It is interesting to see how this results extends to positive supermartingales which aren't potentials

46. THEOREM. <u>Instead of assuming that</u> X <u>is a potential, assume that it is just a positive supermartingale dominated by</u> c, <u>set</u> $X_\infty = \lim_n X_n$ (so that, if X=M-A is DOOB's decomposition for X, we have $M_\infty=A_\infty+X_\infty$). <u>Then</u>

(46.1) $$E[M^p_\infty] \leqq p!c^{p-1}E[X_0]$$

PROOF. Consider the supermartingale $X_0,X_1,\dots,X_n,X_\infty$, 0,0,0,... with respect to $\underline{\underline{F}}_0,\dots,\underline{\underline{F}}_n,\underline{\underline{F}},\underline{\underline{F}},\underline{\underline{F}},\dots$, and apply to it the preceding theorem. The corresponding " A_∞" is $A_n+E[X_n-X_\infty|\underline{\underline{F}}_n]+X_\infty \geqq A_n+X_\infty$. Then let $n\to\infty$.

<u>APPLICATION : BURKHOLDER'S MAXIMAL LEMMA FOR TRANSFORMS</u>

We are going to prove now one of the basic inequalities of BURKHOLDER. The proof rests only on the estimate (46.1) for p=2, and is due to NEVEU . It was rediscovered independently by M.RAO, who first acquainted us with it (though NEVEU's proof was earlier).

47. THEOREM. <u>Let</u> X <u>be a martingale</u>, V <u>a predictable process bounded by</u> 1 <u>in absolute value</u>. <u>Then for every</u> c>0

(47.1) $$cP\{ \sup_n |(V\cdot X)_n| > c \} \leqq 18\|X\|_1$$

18 <u>can be replaced by</u> 9 <u>if</u> X <u>is positive</u>.

PROOF. If Y is a process, we shall set $Y^*_n = \sup_{k\leqq n} |Y_k|$, and $Y^* = Y^*_\infty$. We shall assume first that X is positive, and set $Z=X\wedge c$, a positive supermartingale ; we denote by Z=M-A its DOOB decomposition.

On the set $\{X_n^* \leqq c\}$, the processes X and Z have the same sample path up to time n, and the same is true for V•X and V•Z . Therefore

$$cP\{(V\cdot X)_n^* > c\} \leqq cP\{X_n^* > c\} + cP\{(V\cdot Z)_n^* > c\}$$

The first term to the right can be replaced by $E[X_0]$ (n°5). In the last term, we write V•Z=V•M−V•A . Since V is dominated by 1, A increasing, we have $|V\cdot A| \leqq A$, hence $|V\cdot Z| \leqq |V\cdot M|+A$. This process is a submartingale, and the same is true for its square. We apply (5.6) :

$$cP\{(V\cdot Z)_n^* > c\} \leqq cP\{((|V\cdot M|+A)^2)_n^* > c^2\} \leqq c.\frac{1}{c^2}E[(|V\cdot M|+A)_n^2]$$

We apply the inequality $(u+v)^2 \leqq 2(u^2+v^2)$, and remark that $E[(V\cdot M)_n^2] = E[V_0^2M_0^2+..+V_n^2(M_n-M_{n-1})^2] \leqq E[M_0^2+..+(M_n-M_{n-1})^2] = E[M_n^2]$. Therefore

$$cP\{(V\cdot X)_n^* > c\} \leqq E[X_0] + \frac{2}{c}E[M_n^2+A_n^2]$$

Since X=M−A is positive, A_n is smaller than M_n. On the other hand, (46.1) gives us $E[M_n^2] \leqq 2cE[Z_0] \leqq 2cE[X_0]$, and we get (47.1) with 9 instead of 18.

To deal with the general case, we represent X as a difference X'−X" of two positive martingales (n°16 : (47.1) is trivial if X isn't bounded in L^1). Then

$$P\{(V\cdot X)^* > c\} \leqq P\{(V\cdot X')^* > \frac{c}{2}\} + P\{(V\cdot X")> \frac{c}{2}\}$$

The left member of (47.1) thus is dominated by $18(\|X'\|_1+\|X"\|_1)$. This can be taken to be $18\|X\|_1$ (n°16).

We can now deduce from the MARCINKIEWICZ interpolation theorem and the maximal lemma (47.1) the general BURKHOLDER inequalities for transforms :

48. THEOREM. Let X be a martingale, V a predictable process bounded by 1 in absolute value. Then for $1<p<\infty$.

(48.1) $$\|V\cdot X\|_p \leqq c_p\|X\|_p$$

the finite constant c_p depending only on p.

PROOF. We fix n and define the following operator on $L^1(\underline{\underline{F}})$: to $Y \in L^1(\underline{\underline{F}})$ we associate the martingale $(E[Y|\underline{\underline{F}}_n]) = \overline{Y}$, then the transform $V\cdot\overline{Y}$, then finally the random variable $(V\cdot\overline{Y})_n = Y'_n$. The operator $Y \mapsto Y'_n$ is linear, and satisfies the properties

$cP\{|Y'_n|>c\} \leq A\|Y\|_1$ (A=18 : this is a " weak 1-1 inequality" , see n°7)

$\|Y'_n\|_2 \leq B\|Y\|_2$ (B=1)

The MARCINKIEWICZ theorem[1] then tells us that for $1<p\leq 2$ a constant c_p exists such that

(48.2) $\|Y'_n\|_p \leq c_p\|Y\|_p$

and we get (48.1) when $n\to\infty$, since c_p depends only on p, not on n . To get the result for $2<p<\infty$, we denote by q the exponent conjugate to p, which belongs to]1,2[. Let Z be a second random variable in $L^1(\underline{\underline{F}})$, to which we associate the martingale $\overline{Z}$ and the random variable Z'_n as above. We have from HÖLDER's inequality and the preceding result

(48.3) $E[|Y.Z'_n|] \leq \|Y\|_p\|Z'_n\|_q \leq c_q\|Y\|_p\|Z\|_q$

Assume first that Z is bounded. Then the following computation makes sense :

$$E[Y.Z'_n] = E[\,V_0\overline{Z}_0Y+\ldots+V_n(\overline{Z}_n-\overline{Z}_{n-1})Y]$$

$$= E[\,V_0\overline{Z}_0\overline{Y}_0+\ldots+V_n(\overline{Z}_n-\overline{Z}_{n-1})(\overline{Y}_n-\overline{Y}_{n-1})]$$

since $Y-\overline{Y}_n$ is orthogonal to $\underline{\underline{F}}_n$ and $\overline{Z}_n-\overline{Z}_{n-1}$ to $\underline{\underline{F}}_{n-1}$. This is now symmetric in Y and Z, and so $E[Y.Z'_n]$ by $E[Y'_n.Z]$. So (48.3) gives us

$|E[Y'_n.Z]| \leq c_q\|Y\|_p\|Z\|_q$ if Z is bounded

Taking a supremum over all bounded Z with $\|Z\|_q=1$ we get that $\|Y'_n\|_p \leq c_q\|Y\|_p$, and we let again $n\to\infty$.

ESTIMATING DOOB'S DECOMPOSITION : GENERAL CASE

49 . We assume again that X is a potential, generated by the predictable increasing process A, with the DOOB decomposition X=M-A given by (45.1) as above. Let B be some other - not necessarily adapted - increasing process such that

(49.1) $X_n = E[B_\infty - B_n|\underline{\underline{F}}_n]$ for every n

Since $B_0=0$, B_∞ is integrable. We have for every stopping time T $X_T = E[B_\infty - B_T|\underline{\underline{F}}_T]$ a.s.

Let us choose for T the stopping time $\inf\{n : A_{n+1}>c\}$. Remark

[1] See references to n°7.

that $A_T \leq c$ (even on $\{T=0\}$, since $A_0=0$) and that $\{A_\infty > c\} = \{T<\infty\}$ belongs to $\underline{\underline{F}}_T$. The relation $X_T = E[A_\infty | \underline{\underline{F}}_T] - A_T$ gives us the basic chain of inequalities

$$\int_{\{A_\infty > c\}} (A_\infty - c) \leq \int_{\{A_\infty > c\}} (A_\infty - A_T) = \int_{\{A_\infty > c\}} X_T \qquad (49.1)$$

To estimate the right side, we have two possibilities : either replace X_T by $X^* = \sup_n X_n$; or remark that X_T can be replaced by $B_\infty - B_T$, and then by B_∞ which is larger. We thus get two interesting inequalities

$$(49.2) \qquad \int_{\{A_\infty > c\}} (A_\infty - c) \leq \int_{\{A_\infty > c\}} X^*$$

$$(49.3) \qquad \int_{\{A_\infty > c\}} (A_\infty - c) \leq \int_{\{A_\infty > c\}} B_\infty$$

from which we shall deduce some consequences.

50. The first one is RAO's remark , which is basic in his proof of the DOOB decomposition for continuous time potentials (see RAO's paper [23]) . Assume we have a sequence of potentials X^n , and consider the corresponding increasing processes A^n . We say that the potentials X^n belong uniformly to the class (D) if all random variables X^n_S ($n \in \underset{\sim}{N}$, S stopping time) are uniformly integrable. Then RAO's remark is the following : all random variables A^n_∞ are uniformly integrable.

This is quite easy : on $\{A_\infty > 2c\}$ we have $A_\infty - c \geq A_\infty/2$. Therefore we have from (49.1)

$$\int_{\{A_\infty > 2c\}} A_\infty \leq 2 \int_{\{A_\infty > c\}} X_T \qquad (50.1)$$

We insert indices n, and wonder whether the left side is uniformly small for c large enough. Since all random variables $X^n_{T_n}$ are uniformly integrable, is suffices ([1], chapter I, T19) to show that $P\{A^n_\infty > c\}$ is uniformly small. Now this is smaller than $\frac{1}{c}E[A^n_\infty] = \frac{1}{c}E[X^n_0]$, and $E[X^n_0]$ is bounded in n .

A THEOREM OF BURKHOLDER-DAVIS-GUNDY

Coming back to the notations of n°49, we may replace in (50.1) X_T by X^* or B_∞, thus getting " maximal lemmas"

$$2cP\{A_\infty>2c\} \leq 2\int_{\{A_\infty>c\}} X^* \qquad (\text{ or } B_\infty)$$

This is quite similar to (9.1) , and almost the same proof as in n°9 would lead to L^p norm inequalities between X^* , B_∞, A_∞ - this was a nice remark of KESTEN on RAO's paper [23]. But NEVEU remarked that much more could be deduced from the inequalities (49.2) and (49.3) themselves . We are going to follow his proof. We need some preliminaries on convex functions.

51 . Let F be a convex function on $[0,\infty[$, increasing, such that $F(0)=0$, $\lim_{t\to\infty} \frac{F(t)}{t} = \infty$[1]. We know that F has at every point t a right derivative f(t) and a left derivative f(t-) : f is an increasing, right continuous function. We say that F is <u>tame</u>[1] if it satisfies a growth condition

(51.1) $\qquad F(2t)\leq AF(t)$ for all t

For instance, $F(t)=t^p$, $F(t)=(t+1)\log(t+1)$ are tame, while $F(t)=e^t$ isn't. If F is tame, we have

$$f(2t) \leq \frac{1}{2t}\int_{2t}^{4t} f(s)ds \leq \frac{F(4t)}{2t} \leq A\frac{2F(t)}{2t} \leq \frac{A^2}{2}f(t)$$

Therefore, f also satisfies a growth condition with some other constant a

(51.2) $\qquad f(2t) \leq af(t)$

Conversely, it is very easy to see that (51.2) implies (51.1) with A=2a .

Remark now that f(t) tends to infinity with t, and consider its inverse function

(51.3) $\qquad g(t)= \inf\{ s : f(s)>t \}$

This is a right continuous , increasing function. Let us set

(51.4) $\qquad G(t) = \int_0^t g(s)ds$

This is again a convex, increasing function with G(0)=0, the function <u>conjugate</u> to F : a small drawing of the graph of f will convince the reader of the validity of YOUNG's inequality :

[1]We exclude the functions with linear growth, and point out explicitly below the results which are valid for F(t)=t .

(51.5) $uv \leqq F(u)+G(v)$

On the other hand, we have

(51.6) $g(f(t)) \geqq t \quad , \quad g^{-}(f^{-}(t)) \leqq t$

where f^{-}, g^{-} denote the mappings $t \mapsto f(t-)$, $t \mapsto g(t-)$. Following NEVEU, we prove a lemma :

LEMMA . Assume F is tame. There then exist constants $v>u>0$ (depending only on a in (51.2)) such that

(51.7) $G\left(\frac{f(t)}{u}\right) \leqq \frac{F(t)}{v}$

PROOF. Let m be an integer, to be chosen later. We write that $G(s) \leqq sg^{-}(s)$, taking s to be $f^{-}(t)/a^{m}$

$$G\left(\frac{f^{-}(t)}{a^{m}}\right) \leqq \frac{f^{-}(t)}{a^{m}} g^{-}\left(\frac{f^{-}(t)}{a^{m}}\right) \leqq \frac{f^{-}(t)}{a^{m}} g^{-}\left(f^{-}\left(\frac{t}{2^{m}}\right)\right)$$

(we have used (51.2)). Taking now (51.6) into account , we get

$$G\left(\frac{f^{-}(t)}{a^{m}}\right) \leqq \frac{tf^{-}(t)}{2^{m}a^{m}}$$

On the other hand, $tf^{-}(t) \leqq 2a\frac{t}{2}f^{-}(\frac{t}{2}) \leqq 2aF(t)$. Therefore

$$G\left(\frac{f^{-}(t)}{a^{m}}\right) \leqq \frac{F(t)}{(2a)^{m-1}}$$

We can replace f^{-} by f, since G and F are continuous. We just set now $u=a^{m}$, $v=(2a)^{m-1}$, m being large enough so that $v>u$.

We are now ready to prove NEVEU's result, which is similar to theorem 9 :

52. THEOREM. Let U and V be two positive random variables such that

(52.1) $\int_{\{U>c\}} (U-c) \leqq \int_{\{U>c\}} V$

Let F be tame as above. There then exists a constant K (depending only on the constant a in (51.2)) such that

(52.2) $E[F \circ U] \leqq KE[F \circ V]$

In particular, taking $F(t)=t^{p}$

(52.3) $\|U\|_{p} \leqq p\|V\|_{p}$ (valid also for $p=1$)

PROOF. We have $F(t)=\int_0^t f^-(s)ds = \int_0^t ds \int_{[0,s[} df(c)$ (df has a mass at 0 equal to f(0)). This is also $\int_{[0,t[} df(c)\int_c^t ds = \int_{[0,t[}(t-c)df(c)$. Therefore

$$E[F\circ U]=E[\int_{[0,U[}(U-c)df(c)\ ] = \int_{[0,\infty[} df(c)\int_{\{U>c\}}(U-c)dP$$

We apply (52.1)

$$E[F\circ U] \leq \int_{[0,\infty[} df(c)\int_{\{U>c\}} VdP = E[V\int_{[0,U[} df(c)]=E[V.f^-(U)]$$

In the L^p case, this is just $E[U^p]\leq pE[V.U^{p-1}]$, and (52.3) follows from an easy application of HÖLDER's inequality as in n°9. In the general case, we apply YOUNG's inequality : $st \leq F(s)+G(t)$, with $s=V$, $t=f^-(U)/u$ (the constants u and v are those of (51.7)).

$$E[V.f^-\circ U/u] \leq E[\ F\circ V + G(\frac{f^-\circ U}{u})]\leq E[F\circ V]+ \frac{1}{v}\ E[F\circ U]$$

according to (51.7). Finally, multiplying by u

$$E[F\circ U] \leq uE[F\circ V] + \frac{u}{v}E[F\circ U]$$

Now u/v is <1, and we get (52.2) with $K= \frac{uv}{v-u}$, provided we know that $E[F\circ U]$ is finite. The general case requires a truncation argument (left to the reader) as in n°9.

We state explicitly the results which follow from the application of theorem 52 to (49.2) and (49.3). **There is an error in the numbering of theorems !**

52'. THEOREM. Let X be a positive supermartingale, X=M-A be its DOOB decomposition . Then if F is tame

(52'.1) $\qquad E[F\circ M_\infty\ ] \leq KE[F\circ X^*]$ [1] $\qquad (X^* = \sup_n X_n\ )$

If $1<p<\infty$, we have

(52'.2) $\qquad \|X^*\|_p \leq q\|M\|_p$ [2] $\quad , \quad \|M\|_p\leq p\|X^*\|_p$ [1]

PROOF. If X is a potential, $M_\infty=A_\infty$ and (52'.1), as well as the second inequality in (52'.2), have just been proved. The general case follows as in n°46. The first inequality in (52'.2) isn't new : it is just (8.1), X being dominated by M .

From (49.3), we get the theorem of BURKHOLDER, DAVIS, GUNDY :

---

[1] valid (trivial) for F(t)=t, p=1 . [2] false for p=1

53. THEOREM . Let B be a (not necessarily adapted) increasing process with $B_\infty \in L^1$, and let A be the predictable increasing process associated to it :

$$A_0=0=B_0 \quad , \quad A_{n+1}-A_n = E[B_{n+1}-B_n|\underline{\underline{F}}_n]$$

Then if F is tame[1]

(53.1) $E[F\circ A_\infty] \leqq KE[F\circ B_\infty]$ ($\|A_\infty\|_p \leqq p\|B_\infty\|_p$, $1\leqq p<\infty$)

Conversely, if B is adapted and $B_{n+1}-B_n \leqq 1$ for all n ,

(53.2) $E[F\circ B_\infty] \leqq KE[F\circ(A_\infty+1)]$ ($\|B_\infty\|_p \leqq p(1+\|A_\infty\|_p)$) .

PROOF. (53.1) is just (49.3) plus theorem 52. To prove (53.2), set

$$T = \inf \{ n : B_n > c \}$$

Then $B_T \leqq c+1$, and the same proof as for (49.3) gives

$$\int_{\{B_\infty>c\}} (B_\infty-c-1) \leqq \int_{\{B_\infty>c\}} A_\infty$$

We take 1 to the right side and apply n°52.

QUADRATIC VARIATION AND VARIANCE PROCESS

54. Let X be any process. We define a new process [X,X] as

(54.1) $[X,X]_0=X_0^2$, $[X,X]_n = X_0^2+(X_1-X_0)^2+..+(X_n-X_{n-1})^2$

More generally, if X and Y are two processes, we may define

(54.2) $[X,Y]_0=X_0Y_0$, $[X,Y]_n = X_0Y_0+..+(X_n-X_{n-1})(Y_n-Y_{n-1})$

so that $[X,Y]=\frac{1}{2}([X+Y,X+Y]-[X,X]-[Y,Y])$: these definitions, and the following ones, will be basic for the theory of stochastic integrals, and we shall extend them to continuous time (at least for suitable processes).

The process [X,X] is increasing[2] The quadratic variation of X is the increasing process $Q_n^X = \sqrt{[X,X]_n}$. On the other hand, if the random variables X_n belong to L^2 , we may define a new process $\langle X,X\rangle$ as

(54.3) $\langle X,X\rangle_0=E[X_0^2|\underline{\underline{F}}_0]$, $\langle X,X\rangle_n=E[X_0^2|\underline{\underline{F}}_0[+..+E[(X_n-X_{n-1})^2|\underline{\underline{F}}_{n-1}]$

[1] (53.1) is trivially true for F(t)=t

[2] From a strictly technical point of view, it isn't an "increasing process", however, unless $X_0=0$.

This is a predictable process. The processes $[X,X]$ and $\langle X,X\rangle$ are associated. More generally, one may define

(54.6) $\langle X,Y\rangle_n = E[X_0Y_0|\underline{\underline{F}}_0] +\ldots+E[(X_n-X_{n-1})(Y_n-Y_{n-1})|\underline{\underline{F}}_{n-1}]$

In the particular case that $X_n=x_0+\ldots+x_n$, where x_i denotes a random variable with $E[x_i]=0$, $E[x_i^2]<\infty$, independent of $\underline{\underline{F}}_{i-1}$ (x_0 also independent of $\underline{\underline{F}}_0$), we have $\langle X,X\rangle_n= \sigma_0^2+..+\sigma_n^2$, whence the convenient name of <u>variance process</u> for the process $\langle X,X\rangle$, and notation σ_n^X for $\langle X,X\rangle_n^{1/2}$. We are going to prove some inequalities on the processes Q^X and σ^X, but let us just , however, <u>state</u> the deepest result concerning Q^X , which is due to BURKHOLDER, DAVIS and GUNDY (see [24]) :

55 . THEOREM . <u>Let</u> X <u>be a martingale</u> , <u>and</u> F <u>be tame</u> (<u>n°51</u>). <u>Set</u> $X^* = \sup_n |X_n|$. <u>Then</u>

(55.1) $CE[F\circ Q_\infty^X] \leq E[F\circ X^*] \leq C'E[F\circ Q_\infty^X]$ (<u>also true for</u> $F(t)=t$)

<u>The constants</u> C <u>and</u> C', <u>both</u> >0 <u>and finite</u>, <u>depend only on the growth constant</u> (51.1).

Before we state the second part, let us make a remark : in our definition of tame functions (51) we have found it convenient to assume that $F(t)/t \underset{t\to\infty}{\longrightarrow} \infty$: convex functions such that $F(t)/t$ is bounded have linear growth, and theorem 52 is trivial for them. On the other hand, theorem 55 is true, <u>and non trivial</u>, for $F(t)=t$ (DAVIS [25]).

56. THEOREM. <u>Assume that</u> $\sup_n |X_{n+1}-X_n| \leq 1$, <u>and set</u> $\sigma_\infty^X=\langle X,X\rangle_\infty^{1/2}$

<u>Then</u>

(56.1) $KE[F\circ(1+\sigma_\infty^X)] \leq E[F\circ(1+Q_\infty^X)] \leq K'E[F\circ(1+\sigma_\infty^X)]$

Here too the result is true for $F(t)=t$. We aren't going to prove these results, because the available proofs are probably not yet " ripe" , and hence not quite suitable for a book. Let us just indicate some consequences. The first corollary concerns martingale transforms : the fact that $[V\cdot X, V\cdot X]=V^2\cdot[X,X]$ gives at once the nice result that

57 . THEOREM. Let X be a martingale, and V a predictable process bounded by 1 in absolute value. Then if F is tame (also if $F(t)=t$)

(57.1) $E[\,F\circ(\sup_n |(V\cdot X)_n|)] \leqq CE[\,F\circ(\sup_n |X_n|\,)]$

A consequence of 56, now : it improves on (53.2) in the fact that tame functions are replaced by their square roots. There probably exists a direct proof, and it is safe to conjecture that, here too, results of this kind will "ripen".

58 . THEOREM . Let B be an (adapted) increasing process. We set as usual $B_n=b_0+\dots+b_n$, $a_0=0$, $a_n=E[b_n|\underline{\underline{F}}_{n-1}]$, $A_n=a_0+\dots+a_n$, and assume that $b_n\leqq 1$ for all n. Then if F is tame (also for $F(t)=t$)

(58.1) $KE[F\circ(1+A_\infty^{1/2})]\leqq E[F\circ(1+B_\infty^{1/2})]\leqq K'E[F\circ(1+A_\infty^{1/2})]$

PROOF. Adjoin to the space a sequence of BERNOULLI random variables r_n , independent from $\underline{\underline{F}}$ ($P(r_n=\pm1)=\frac{1}{2}$) . Let $\underline{\underline{G}}_n$ be generated by $\underline{\underline{F}}_n$ and the random variables $r_0,\dots,r_n$. Set $x_n=r_n\sqrt{b_n}$, so that $E[x_{n+1}|\underline{\underline{G}}_n] = E[x_{n+1}|\underline{\underline{G}}_n\vee\underline{\underline{F}}_{n+1}|\underline{\underline{G}}_n]=0$, and the process $X_n=x_0+\dots+x_n$ is a martingale. We remark that

$$[X,X]_n = B_n \quad , \quad <X,X>_n = A_n$$

and we apply (56.1) .

QUADRATIC VARIATION

We start with some remarks : may be they are interesting !

57. Let X be a martingale bounded in L^2. Set $X_\infty = \lim_n X_n$ and

(57.1) $Y_n = E[X_\infty^2|\underline{\underline{F}}_n] - X_n^2 + (X_n-X_{n-1})^2$; $Y_0=E[X_\infty^2|\underline{\underline{F}}_0]$

(57.2) $Y'_n = E[X_\infty^2|\underline{\underline{F}}_n] - X_n^2$ $(n>0)$; $Y'_0=E[X_\infty^2|\underline{\underline{F}}_0]$.

We have $E[X_\infty^2|\underline{\underline{F}}_n] \geqq (E[X_\infty|\underline{\underline{F}}_n])^2= X_n^2$ (JENSEN 's inequality). Therefore Y' is positive, and the same is true for Y, which is larger . X_n tends to 0 in L^2, hence X_n^2 and $E[X_\infty^2|\underline{\underline{F}}_n]$ tend to X_∞^2 in L^1 as $n\to\infty$, while $(X_n-X_{n-1})^2$ tends to 0 in L^1, so that Y_n and Y'_n tend to 0 in L^1 as $n\to\infty$

An easy computation now gives

$$E[Y_0-Y_1|\underline{\underline{F}}_0] = X_0^2 \quad , \; E[Y_n-Y_{n+1}|\underline{\underline{F}}_n] = (X_n-X_{n-1})^2$$

Hence Y is a supermartingale (potential) and the predictable increasing process generating it is A given by

(57.3) $A_0=0 \quad , \quad A_n=[X,X]_{n-1}$ for $n>0$.

Similarily ,

$$E[Y'_0-Y'_1|\underline{\underline{F}}_0]=E[X_1^2|\underline{\underline{F}}_0] = E[X_0^2 + (X_1-X_0)^2|\underline{\underline{F}}_0]$$

$$E[Y'_n-Y'_{n+1}|\underline{\underline{F}}_n] = E[(X_{n+1}-X_n)^2|\underline{\underline{F}}_n]$$

Y' is also a potential, generated by A' :

(57.4) $A'_0=0 \quad , \quad A'_n= \langle X,X\rangle_n$ for $n>0$

We may now apply the results concerning DOOB's decomposition : first

58. THEOREM. Let the martingale X be bounded by c in absolute value. Then for any integer $p \geqq 1$

(58.1) $$E[\,[X,X]_\infty^p\,] \leqq p!(4c^2)^{p-1}E[X_\infty^2\,] \leqq p!(2c)^{2p}$$

(58.2) $$E[\,\langle X,X\rangle_\infty^p] \leqq p!c^{2(p-1)}E[X_\infty^2\,] \leqq p!c^{2p}$$

PROOF. If X is bounded by c, Y' is bounded by c^2 and Y by $4c^2$. One then applies theorem 45.

Next, we may apply theorem 52 and get that, if X is bounded in L^p ($p>2$), then $[X,X]_\infty$ and $\langle X,X\rangle_\infty$ belong to $L^{p/2}$, and therefore Q_∞^X and σ_∞^X to L^p - but this isn't interesting[1], since these properties (at least those concerning Q^X) are true for all $p>1$, and are in fact norm equivalences in L^p : we are going now to see it.

QUADRATIC VARIATION : THE BURKHOLDER L^p INEQUALITIES

59 . We are going to use the classical proof using the BURKHOLDER L^p inequalities and the RADEMACHER functions. For our purposes, we just need to know they are functions r_n ($n \geqq 0$) on the unit interval I (with BOREL σ-field and usual measure) which on this space are BERNOULLI random variables : independent, distributed with mass 1/2 at ±1 . We need KHINCHIN's lemma :

---

1 See however the end of appendix 2 (GARSIA's remark).

THEOREM. Let (a_n) be a sequence of real numbers, such that $\sum_n a_n^2 < \infty$, and let f be $\sum_n a_n r_n$ (a series which obviously converges in L^2, and may be shown to converge a.s.). There then exist finite, non zero constants A_p, B_p $(1 \leqq p < \infty)$ such that

$$A_p.(\sum a_n^2)^{1/2} \leqq \|f\|_p \leqq B_p.(\sum a_n^2)^{1/2} .$$

PROOF. a) Existence of B_p : since $\|f\|_p$ increases with p, it suffices to consider the case of an integer p . Since the r_n are independent random variables, we have

$$E[e^f] = \prod_n E[e^{a_n r_n}] = \prod_n \operatorname{Cosh}(a_n) \leqq \exp(\sum_n a_n^2)$$

Hence $E[e^{|f|}] \leqq 2\exp(\sum a_n^2)$. On the other hand, the left side is larger than $E[|f|^p/p!]^n$. Finally, $\sum a_n^2 \leqq 1$ implies $\|f\|_p \leqq (2ep!)^{1/p}$, the desired result.

b) Existence of A_p : everything is trivial if $p \geqq 2$, since $\|f\|_p$ increases with p and $\sum a_n^2$ is $\|f\|_2^2$; we just take $A_p=1$. To deal with the case $1 \leqq p < 2$, we may assume that $\sum a_n^2 = 1$, and we recall that the function $t \mapsto \log \|f\|_{1/t}$ is convex in $[1,\infty[$. Since 2 lies between p and 4, we may write

$$\frac{1}{2} = \frac{r}{p} + \frac{s}{4} \quad (r \geqq 0,\ s \geqq 0,\ r+s = 1)$$

Then we have $1=\|f\|_2 \leqq \|f\|_p^r \|f\|_4^s \leqq \|f\|_p (B_4)^s$, and we may take $A_p = B_4^{-s}$.

The empty space on this page shows clearly that a proof has been shortened : we have borrowed the nice proof of a) given in STEIN's book (Singular integrals and differentiability properties of functions, Princeton Math. series, 1970).

60. THEOREM. Let X be a martingale, and p belong to $]1,\infty[$. Then we have

(60.1) $$C\| Q^X_\infty \|_p \leqq \| X \|_p \leqq C' \| Q^X_\infty \|_p$$

where C and C' are finite, non zero, and depend only on p.

(Taking theorem 8 into account, this appears as a special case of theorem 55).

PROOF. For fixed t, consider the process V^t (a deterministic process !) defined by $V^t_n = r_n(t)$, and the transform $V^t.X$. We have from BURKHOLDER's inequalities for transforms (th.48)[1]

$$E[|(V^t.X)_n|^p] \sim \|X\|^p_p$$

the symbol $\sim$ meaning that the ratio of the right side to the left lies between two finite, non zero constants depending only on p. Since they do not depend on t, we have the same result for the integral of the left side over t. This is

$$\int (|\sum_{k=0}^n r_k(t)x_k(\omega)|)^p \, dP(\omega)dt$$

We integrate in t first, and use the fact that

$$\int(|\sum_{k=0}^n r_k(t)x_k(\omega)|)^p dt \sim ((\sum_0^n x^2_k(\omega))^{1/2})^p = (Q^X_n(\omega))^p$$

(theorem 59). We finish the proof by integrating over ω , and then letting $n \to \infty$.

For p=1 , we may prove the following : if $Q^X_\infty \in L^1$, then $\int E[|\sum_{k=0}^n r_k(t)x_k(\omega)|]dt \leqq cE[Q^X_\infty]$. Therefore, some t must exist such that

$$E[|(V^t.X)_n|] \leqq cE[Q^X_\infty] \text{ for all } n$$

Otherwise stated, $V^t.X$ is bounded in L^1, and BURKHOLDER's maximal lemma gives us (since $X= V^t.(V^t.X)$; see n°47)

(60.2) $$\lambda P\{ \sup_n |X_n|>\lambda \} \leqq c' \| Q^X_\infty\|_1$$

However, this is uninteresting : the theorem of BURKHOLDER-DAVIS-GUNDY quoted in n°55 tells us that $\sup_n|X_n|$ belongs to L^1 if Q^X_∞ does ! The following inequality

(60.3) $$\lambda P\{ Q^X_\infty >\lambda \} \leqq c'' \| X \|_1$$

---

[1] X is also a transform of $V^t.X$, since $r_k(t)=\pm 1$! $X=V^t.(V^t.X)$.

is more interesting, and can be proved by methods similar to those of n° 47 .

INEQUALITIES CONNECTED WITH THE VARIANCE PROCESS

The results concerning the variance process cannot be so sharp as those concerning quadratic variation. Consider for instance the case of a martingale X with independent increments . If the increments do not belong to L^2, then the variance process doesn't exist at all. If they do, then the only significant property is whether the constant $\langle X,X\rangle_\infty$ is finite or not - there cannot exist anything like L^p bounds, etc .

We start with a very nice inequality due to DUBINS and FREEDMAN [26]. Let a be a number >0, and q_a be the function on $\mathbb{R}$

(61.1) $\quad q_a(t)=1$ for $t\geqq 0$, $q_a(t) = \frac{1}{1-at}$ for $t<0$

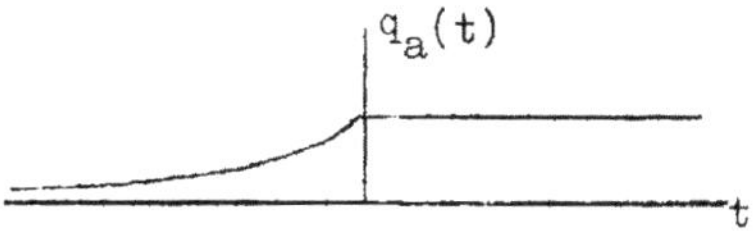

61. LEMMA. Let Θ be a probability law on $\mathbb{R}$ with finite second moment, mean $-m < 0$, variance σ^2 such that

(61.2) $\quad a\sigma^2 \leqq m$

Then we have for all x

(61.3) $\quad \int q_a(x+y)\Theta(dy) \leqq q_a(x)$.

PROOF[1]. We are going to prove the following, slightly more precise result : if X is a random variable on some probability space, with mean 0 and variance σ^2 , then we have for all μ

(61.4) $\quad E[q_a(\mu+X)] \leqq q_a(\mu+a\sigma^2)$

(to get (61.3) take Θ as probability law on the line, $\mu=x-m$, and $X(t)=t+m$; since q_a is increasing we have $q_a(x-m+a\sigma^2) \leqq q_a(x)$).

We reduce to $a=1$, and set $q=q_1$. Since $q\leq 1$, there is nothing to prove if $\mu\geqq 0$, so we assume $\mu<0$. Consider the parabolic graph of the following function r

$$r(\mu+u) = q(\mu) + (u+u^2)q'(\mu) .$$

1. This proof was communicated by J.L.DOOB.

(the reason for considering r is that $r(\mu)=q(\mu)$, $r'(\mu)=q'(\mu)$, and $r(0)=1=q(0)$). It is very easy to check that

$$r(t)-q(t) = [\,1-\mu+(t-\mu)+(t-\mu)^2]\frac{1}{(1-\mu)^2} - \frac{1}{1-t}$$
$$= \frac{-t(t-\mu)^2}{(1-t)(1-\mu)^2}$$

is positive on the left half line, and on the right half line we have $r(t)\geqq r(0)=1=q(t)$. Therefore

$$E[q(\mu+X)] \leqq E[r(\mu+X)] = q(\mu) + \sigma^2 q'(\mu) \leqq q(\mu+\sigma^2)$$

(this last inequality by convexity if $\mu+\sigma^2\leqq 0$, the whole thing being trivial if $\mu+\sigma^2 \geqq 0$) . The lemma is proved.

REMARKS. Let Y be any random variable with a finite second moment and mean $m\leqq 0$. Set $s^2 =E[Y^2]$. Then $E[q_a(\mu+Y)]\leqq q_a(\mu+as^2)$. Apply the preceding result to Y-m and μ+m ; it yields for the left side the estimate $E[q_a(\mu+Y)]\leqq q_a(\mu+m+a\sigma^2)$. Now $\mu+m\leqq\mu$, and $\sigma^2\leqq s^2$.

The reader probably would like to know something about the motivations of the preceding lemma. Those of DUBINS from gambling theory, though quite natural, would take us too far afield. But DOOB communicated us some very interesting remarks on this subject. Let us say that a function f - say, positive and increasing, but these properties can be weakened - is <u>admissible</u> if for any random variable X with mean μ and variance σ^2

$$E[f\circ X] \leqq f(\mu+\sigma^2)$$

Then it may be shown (just by considering r.v. X with a two point distribution at 0 and -a, a<0 , but DOOB's proof isn't trivial) that the function q_1 of DUBINS is <u>the smallest admissible</u> f <u>such that</u> f(0)=1 .

62. THEOREM. <u>Let</u> X <u>be a (super)martingale such that</u> $X_0=0$, $X_n\in L^2$ <u>for all</u> n. <u>Let</u> $\langle X,X\rangle$ <u>be its " variance process"</u>

(62.1) $\quad \langle X,X\rangle_0=0 \quad , \langle X,X\rangle_n = \sum_1^n E[x_n^2|\underline{\underline{F}}_{n-1}]$

(x_n <u>denoting</u> X_n-X_{n-1} <u>as usual</u>). <u>Then for any</u> a>0 <u>and any</u> $\underline{\underline{F}}_0$<u>-measurable random variable</u> H_0, <u>the process</u> $(q_a(H_0+X_n-a\langle X,X\rangle_n))$ <u>is a supermartingale</u>.

PROOF. We consider only the martingale case, leaving the supermartingale case to the reader (and the hint to use the preceding first remark). Set $\sigma_n^2 = E[x_n^2 | \underline{\underline{F}}_{n-1}]$, and call $\Theta_n(\omega,.)$ a conditional distribution of $x_n - a\sigma_n^2$ with respect to $\underline{\underline{F}}_{n-1}$ (which exists since $\mathbb{R}$ is a nice Polish space). Its mean is $-a\sigma_n^2$ and its variance σ_n^2 (a.s.). Setting $Z_n = H_0 + X_n - a\langle X,X\rangle_n$ we have

$$E[q_a(Z_n)|\underline{\underline{F}}_{n-1}] = E[q_a(Z_{n-1}+x_n-a\sigma_n^2)|\underline{\underline{F}}_{n-1}] = \int q_a(Z_{n-1}(\omega)+y)\Theta_n(\omega,dy)$$
$$\leqq q_a(Z_{n-1}) \quad \text{a.s.}$$

which is precisely the supermartingale property.

63. COROLLARY . Let X be a (super)martingale such that $X_0=0$, and $X_n \in L^2$ for all n. Let a,b be two numbers >0. Then

(63.1) $\qquad P\{ X_n \geqq a\langle X,X\rangle_n + b$ for some $n \} \leqq \frac{1}{1+ab}$.

(it is shown in [26] that the bound is sharp).

PROOF. We apply the preceding result with $H_0=-b$. The event (63.1) then can be written $P\{q_a(Z_n)\geqq 1$ for some $n\}$, probability smaller than $E[q_a(Z_0)] = q_a(-b) = 1/(1+ab)$, according to n°5, and the fact that $q_a \circ Z_n$ is a positive supermartingale.

We now give two applications , both from [26]. The first one, in the case of a sequence of independent random variables with mean 0 and finite second moments , reduces to the strong law of large numbers. X here is a martingale, not a supermartingale.

64. THEOREM. Under the same hypotheses , $\frac{X_n}{\langle X,X\rangle_n}$ converges a.s. to a finite limit as $n\to\infty$, on the set $\{\langle X,X\rangle_\infty > 0\}$. This limit is 0 on $\{\langle X,X\rangle_\infty = \infty\}$.

PROOF. Let T be the stopping time $\inf\{n : \langle X,X\rangle_{n+1} > N\}$; the martingale $(X_{T\wedge n})$ is bounded in L^2, and therefore converges a.s. In particular, X_n converges a.s. on $\{T=\infty\} = \{\langle X,X\rangle_\infty \leqq N\}$. This settles the matter on $\{\langle X,X\rangle_\infty < \infty\}$.

Next, consider the set $A = \{ \langle X,X\rangle_\infty = \infty , \limsup_{n\to\infty} \frac{|X_n|}{\langle X,X\rangle_n} > 2a \}$. We want to prove that A is a null set for every $a>0$. Now, for any $b>0$, A is contained in the set

$$\{ |X_n| > a\langle X,X\rangle_n + b \quad \text{for some } n \}$$

the probability of which is at most $\frac{2}{1+ab}$ (n°63). Letting $b \to \infty$ we find that $P(A)=0$.

65. The next result concerns a situation already seen in I.22 (BOREL-CANTELLI lemma), II.53, and again II.58 . Let (b_n) be a sequence of positive random variables, adapted to the family $(\underline{\underline{F}}_n)$ and belonging to L^2 . Set as usual $a_0=b_0$, $a_n=E[b_n|\underline{\underline{F}}_{n-1}]$ for $n>0$, $A_n=a_0+\ldots+a_n$, $B_n=b_0+\ldots+b_n$.

THEOREM. Set $V_n= b_0^2+\sum_1^n E[b_k^2|\underline{\underline{F}}_{k-1}]$. Then $\frac{B_n-A_n}{V_n}$ converges a.s. to a finite limit on $\{V_\infty>0\}$, and this limit is a.s. 0 on $\{V_\infty=\infty\}$ (the positivity of the b_i isn't used here).

PROOF. Set $X_n=B_n-A_n$, a martingale with $X_0=0$. The corresponding variance process is

$$\langle X,X\rangle_n = V_n - (a_0^2 + \sum_1^n a_k^2)$$

On $\{\langle X,X\rangle_\infty=0\}$, we have $X_n=0$ for all n, and therefore no convergence problem. Assume $\langle X,X\rangle_\infty > 0$: on $\{0<\langle X,X\rangle_\infty<\infty\}$, $X_n/\langle X,X\rangle_n$ has a finite limit (64), and the same is true for X_n/V_n since $\langle X,X\rangle_n/V_n$ has the finite limit $\langle X,X\rangle_\infty/V_\infty$ (note that V_∞ isn't 0 on this set). The limit is 0 on $\{\langle X,X\rangle_\infty<\infty, V_\infty=\infty\}$. Finally, on $\{\langle X,X\rangle_\infty=\infty\}$, $X_n/\langle X,X\rangle_n$ tends to 0 a.s., and the same is true for X_n/V_n , which is smaller in absolute value.

We know from I.22 that if the random variables b_n are uniformly bounded, A_∞ and B_∞ are simultaneously finite or infinite. We shall prove that, under the same hypothesis, A_n/B_n tends to 1 on the infiniteness set. This is a sharpening of the BOREL-CANTELLI lemma.

66. COROLLARY. Assume the random variables b_n are uniformly bounded by a constant d. Then A_n/B_n tends a.s. to 1 on the set $\{A_\infty=\infty\} \underset{a.s.}{=} \{B_\infty=\infty\}$.

PROOF. We have $E[b_k^2|\underline{\underline{F}}_{k-1}] \leqq da_k$. Therefore on $\{A_\infty=\infty\}$,

- either $V_\infty<\infty$, then B_n-A_n tends to a finite limit, and $(B_n-A_n)/A_n$ tends to 0, the desired result ;
- or $V_\infty=\infty$, then $(B_n-A_n)/V_n$ tends to 0, and the same is true for $(B_n-A_n)/dA_n$ which is smaller in absolute value, whence the result follows.

MARTINGALES WITH BOUNDED JUMPS : EXPONENTIAL BOUNDS

67. We study further the inequalities connected with the variance process, restricting ourselves to the case of martingales X (supermartingales) whose increments are bounded (bounded from above) . We simplify somewhat our notation, setting

(67.1) $\quad x_n = X_n - X_{n-1} \ , \ \sigma_n^2 = E[x_n^2 | \underline{\underline{F}}_{n-1}] \ , \ s_n^2 = \sum_1^n \sigma_n^2 = \langle X,X\rangle_n$

We always assume that $X_0=0$. We also need some elementary inequalities : set $\varepsilon(t) = e^t - 1 - t$ on $\mathbb{R}$. The equation $e^t-1-t = at^2$ $(a>0)$ has a root at 0 , and can have at most one root $\neq 0$ (otherwise, according to ROLLE's theorem , $e^t-1=2at$ would have 2 roots $\neq 0$, which is impossible because of convexity). We apply this with $a = \varepsilon(\lambda)/\lambda^2$, in which case the two roots are 0 and λ . An easy study of the sign at 0 and ∞ shows that

(67.2) $$e^t-1-t \leqq \frac{\varepsilon(\lambda)}{\lambda^2} t^2 \ , \ -\infty < t \leqq \lambda \quad (\lambda \in \mathbb{R})$$
$$e^t-1-t \geqq \frac{\varepsilon(-\lambda)}{\lambda^2} t^2 \ , \quad -\lambda \leqq t < +\infty$$

68. THEOREM. <u>Assume</u> X <u>is a supermartingale, with</u>
$X_0=0$, $x_n \leqq 1$ <u>for all</u> n .
<u>Then the process</u>

(68.1) $$Y_n = \frac{e^{\lambda X_n}}{\prod_{1 \leq k \leq n} (1+E[\varepsilon(\lambda x_n) | \underline{\underline{F}}_{n-1}])}$$

<u>is a supermartingale for all</u> $\lambda \in \mathbb{R}_+$. <u>If</u> X <u>is a martingale and</u> $|x_n| \leqq 1$, <u>then</u> Y <u>is a martingale for</u> $\lambda \in \mathbb{R}$.

PROOF. This is really a trivial and uninteresting result ! We remark that Y_{n+1} is $Y_n . y_n$, where

$$y_n = e^{\lambda x_n} / (1+E[\varepsilon(\lambda x_n) | \underline{\underline{F}}_{n-1}])$$

This is a bounded random variable, and we need only show that $E[y_n | \underline{\underline{F}}_{n-1}] \leqq 1$, or =1 in the martingale case. We just remark that $E[e^{\lambda x_n} | \underline{\underline{F}}_{n-1}] = 1 + E[x_n | \underline{\underline{F}}_{n-1}] + E[\varepsilon(\lambda x_n) | \underline{\underline{F}}_{n-1}]$

69. THEOREM. *Assume* X *is a supermartingale, with* $X_0=0$, $x_n \leqq 1$ *for all* n. *Then the processes*

$$(69.1) \qquad \frac{e^{\lambda X_n}}{\prod_{1\leqq k\leqq n}(1+\varepsilon(\lambda)\sigma_k^2)} \quad , \quad e^{\lambda X_n-\varepsilon(\lambda)<X,X>_n}$$

are supermartingales for $\lambda\in\mathbb{R}_+$. *If* X *is a martingale with* $X_0=0$, $|x_n|\leqq 1$, *the same is true for*

$$(69.2) \qquad e^{\lambda X_n-\varepsilon(|\lambda|)<X,X>_n} \quad \text{for } \lambda\in\mathbb{R}$$

Also the following processes are, in this case, submartingales

$$(69.3) \qquad \frac{e^{\lambda X_n}}{\prod_{1\leqq k\leqq n}(1+\varepsilon(-\lambda)\sigma_k^2)} \quad (\lambda\in\mathbb{R}_+)$$

$$(69.4) \qquad \exp\,(\,\lambda X_n - \log(1+\varepsilon(-\lambda))<X,X>_n\,]$$

PROOF. Quite similar to that of the preceding theorem. Set

$$y_n = \frac{e^{\lambda x_n}}{\prod_{1\leqq k\leqq n}(1+\varepsilon(\lambda)\sigma_k^2)} \quad \text{or} \quad \exp(\,\lambda x_n-\varepsilon(\lambda)\sigma_n^2\,)$$

and check whether $E[y_n|\underline{\underline{F}}_{n-1}]\leqq 1$, otherwise stated, whether $E[e^{\lambda x_n}|\underline{\underline{F}}_{n-1}] \leqq 1+\varepsilon(\lambda)\sigma_n^2$ (which itself is $\leqq e^{\varepsilon(\lambda)\sigma_n^2}$, according to the inequality $1+t\leqq e^t$). We write the left side as $E[1+\lambda x_n+\varepsilon(\lambda x_n)|\underline{\underline{F}}_{n-1}]$, and apply (67.2) to see that $\varepsilon(\lambda x_n)\leqq\varepsilon(\lambda)x_n^2$. We get (69.2) by applying (69.1) to $-X$, in the martingale case.

To prove (69.3), consider again

$$y_n = \frac{e^{\lambda x_n}}{\prod_{1\leqq k\leqq n}(1+\varepsilon(-\lambda)\sigma_k^2)}$$

and prove that $E[y_n|\underline{\underline{F}}_{n-1}] \geqq 1$, or that $E[e^{\lambda x_n}|\underline{\underline{F}}_{n-1}] \geqq 1+\varepsilon(-\lambda)\sigma_n^2$. This we do as above, using the second inequality (67.2) instead of the first.

It remains to see that $1+\varepsilon(-\lambda)\sigma_n^2 \geqq \exp(\,\log(1+\varepsilon(-\lambda)).\sigma_n^2\,)$. Since $\sigma_n^2 \leqq 1$, this amounts to the well known convexity inequality $(1+a)^t \leqq 1+at$ for $a\geqq -1$, $0\leqq t\leqq 1$.

REMARK. If $X=(X_t)_{t\in\mathbb{R}_+}$ is a sample continuous martingale in continuous time, with $X_0=0$, and if $<X,X>$ denotes its variance process (to be defined later), then it may be shown that

(69.5) $\exp\,(\lambda X_t - \frac{\lambda^2}{2}<X,X>_t\,)$ is a martingale for $\lambda\in\mathbb{R}$

(69.6) if $<X,X>_\infty$ is bounded by a constant k, then

$$P\{\ \sup_t X_t \geq ck\ \} \leq e^{-c^2k/2} \qquad (c\geq 0)$$

These properties are extremely useful. (69.5) can be interpreted as the limit of the silly, useless martingale of n°68 when the size of the jumps tends to 0 : this point of view is studied in MAISONNEUVE [27]. Our purpose in n°69 was to find simpler, and similar, supermartingales and submartingales in the discrete case. We are going now to study the discrete analogue of (69.6).

70. Let X be a martingale[1] such that $X_0=0$, $|x_n|\leq 1$ for all n, and assume that

(70.1) $$<X,X>_\infty \leq k$$

Set $X^\times = \sup_n X_n$. We want to estimate $P\{X^\times\geq ck\}$ ($c\geq 0$) . This probability is dominated by

$$P\ \{\ \sup_n e^{\lambda X_n-\varepsilon(\lambda)<X,X>_n} \geq e^{\lambda ck-\varepsilon(\lambda)k}\ \}\quad (\lambda\in\mathbb{R}_+)$$

Since the first random variable of the supermartingale (69.2) is 1, we get from n°5 the estimate

(70.2) $P\{X^\times\geq ck\ \} \leq e^{-\lambda ck+\varepsilon(\lambda)k} \qquad (\lambda\in\mathbb{R}_+)$

We now choose λ so as to minimize the right side. This occurs for $\lambda=\log(c+1)$, and we get

(70.3) $$P\{\ X^\times\geq ck\ \} \leq \left[\frac{e^c}{(c+1)^{c+1}}\right]^k$$

This may be a rather good bound (as suggested by the fact that $e^c/(c+1)^{c+1}$ is <1 for all values of c).

---

1 We use only values of $\lambda\geq 0$ in the proof, therefore it works also for a supermartingale.

71. We are going to test a little the strength of (70.3).

W.F.STOUT has proved in [28] the following remarkable law of the iterated logarithm for martingales (a result simplified and slightly improved in [29]) .

THEOREM. Let X be a martingale such that $X_0=0$, $X_n \in L^2$ for all n, and $s_n \to \infty$ a.s. . Set $u_n=\sqrt{2\log\log s_n}$. Then we have

(71.1) $$\limsup_{n\to\infty} \frac{X_n}{s_n u_n} = 1 \quad \text{a.s.}$$

provided there exists a positive predictable process (K_n) such that

(71.2) $$K_n \underset{n\to\infty}{\longrightarrow} 0 \quad , \quad |x_n| \leqq \frac{K_n s_n}{u_n}$$

We shall not try to prove this result, which is far too technical. We just consider the case of a process with bounded jumps, and even in this case do not try to be complete. Here is the first half of the theorem .

If X is a supermartingale, $x_n \leqq 1$ for all n, $s_n \to +\infty$ a.s., then

$$\limsup_{n\to\infty} \frac{X_n}{s_n u_n} \leqq 1$$

PROOF. We may assume that $X_0=0$. Let p denote a number >1, and consider the stopping times

(71.3) $$T_m = \inf \{ n : s^2_{n+1} \geqq p^{2m} \}$$

The condition that $s_n \to \infty$ implies they are all finite. We have $s^2_{T_m+1} \geqq p^{2m}$, $s^2_{T_m} < p^{2m}$, and therefore $p^{2m}-1 \leqq s^2_{T_m} < p^{2m}$, since the jumps of $<X,X>$ are smaller than 1 .

Denote by Z^m the supermartingale X stopped at T_m , and by $Z^\times_m$ the r.v. $\sup_n Z^m_n$. It will be sufficient to prove that for every number a>1

(71.4) $$P\{Z^\times_m \geqq a p^m \sqrt{2\log m} \ \text{ i.o. } \} = 0 \ .$$

Indeed, consider $\omega \in \Omega$, b>1, and assume $X_n(\omega)>bs_n(\omega)u_n(\omega)$ some large n. Choose m such that $T_m(\omega)<n\leqq T_{m+1}(\omega)$: then $s_n u_n \geqq s_{T_m} u_{T_m}$, equivalent to $p^m\sqrt{2\log\log p^m} \sim p^m\sqrt{2\log m}$. On the

other hand, $X_n(\omega) \leqq Z^{\times}_{m+1}(\omega)$ and finally, if a>1 and p>1 are chosen such that ap<b, we have

$$X_n \geqq bs_n u_n \text{ i.o.} \quad \Rightarrow \quad Z^{\times}_{m+1} \geqq ap^{m+1}\sqrt{2\log m} \quad \text{i.o.}$$

So let us prove (71.4). The supermartingale Z^m has jumps bounded by 1, its total variance $<Z^m,Z^m>_\infty$ is at most p^{2m}, therefore (70.3) with $k=p^{2m}$, $c=ap^{-m}\sqrt{2\log m}$ gives us

$$P\{Z^{\times}_m \geqq ap^m \sqrt{2\log m}\} \quad \leqq \quad \left[\frac{e^c}{(c+1)^{c+1}}\right]^{p^{2m}}$$

We need only show that the series to the right converges : that it a little problem in elementary calculus, the result being that its m-th term is equivalent to $e^{-a^2\log m}$, a convergent series if a>1.

72. We now consider a martingale X with jumps $|x_n|\leqq 1$, $\lim_n s_n = \infty$ (and we can assume as above that $X_0=0$). The second half of the theorem, namely that $\limsup_n X_n/s_n u_n \geqq 1$, unfortunately is much more technical than the first one. So we shall just prove that

(72.1) $$\limsup_n \frac{X_n}{s_n u_n} \geqq C \quad \text{a.s.}$$

where C is some constant >0, to be explicitly given later on. This will be a little more complicated than the first part, but still not too expensive for us.

We are going to prove that, for all positive $a<\frac{1}{2}$, we a.s. have

(72.2) $$X_{T_{m+1}} - X_{T_m} \geqq p^m \sqrt{p^2-1}\sqrt{a\log m} \quad \text{i.o.}$$

Since we also know, from the first part of the theorem, that $X_{T_m} \geqq -2p^m\sqrt{2\log m}$ a.s. for all large enough m , we deduce from (72.2) that

(72.3) $$X_{T_{m+1}} \geqq \left(\sqrt{\tfrac{a}{2}}\sqrt{p^2-1} - 2\right) p^m \sqrt{2\log m} \qquad \text{i.o.}$$

On the other hand, we have seen some time ago that $s_{T_{m+1}} u_{T_{m+1}} \sim p^{m+1}\sqrt{2\log m}$. Taking p large we get (72.1) with C = 1/2. If we could take a<2 instead of $a<\frac{1}{2}$, we would get the best possible result (C=1)

Let A_m be the event (72.2) . We want to prove that $\sum I_{A_m} = +\infty$ a.s.. According to LEVY's version of the BOREL-CANTELLI lemma (I.22), this is equivalent to $\sum P(A_m | \underline{\underline{F}}_{T_m}) = +\infty$ a.s. . On the other hand, we may interpret $X_{T_{m+1}} - X_{T_m}$ as Y_∞ for the martingale $Y_k = X_{(T_m+k)\wedge T_{m+1}} - X_{T_m}$, relative to the family $\underline{\underline{G}}_k = \underline{\underline{F}}_{T_m+k}$. This martingale satisfies to the following conditions

$$Y_0 = 0 \ , \ |y_n| \leqq 1 \text{ for all } n \ , \ A \leqq \langle Y,Y\rangle_\infty \leqq B$$

with

(72.4) $\quad A = p^{2(m+1)} - p^{2m} - 1 \ , \quad B = p^{2(m+1)} - p^{2m} + 1$.

The process $\exp(\lambda Y_n - \varepsilon(\lambda)\langle Y,Y\rangle_n)$ being a supermartingale, the first random variable of which is equal to 1, we have

(72.5) $\quad E[e^{\lambda Y_\infty} | \underline{\underline{G}}_0] \leqq e^{B\varepsilon(\lambda)} \qquad (\lambda \in \mathbb{R}_+)$

Similarily, $\exp(\lambda Y_n - \log(1+\varepsilon(-\lambda))\langle Y,Y\rangle_n)$ being a submartingale, we have

(72.6) $\quad E[e^{\lambda Y_\infty} | \underline{\underline{G}}] \geqq e^{A\log(1+\varepsilon(-\lambda))} \qquad (\lambda \in \mathbb{R}_+)$

From these inequalities, we must deduce lower estimates for $P\{Y_\infty \geqq x | \underline{\underline{G}}_0\}$ for large x. This is the place where STOUT (following KOLMOGOROV's method for independent r.v.) becomes very complicated. We shall use something much coarser : the following little lemma (probably due to PALEY-ZYGMUND)

LEMMA. <u>Let</u> Z <u>be a positive random variable with a finite expectation</u>. <u>Assume</u> $E[Z] \geqq u$, $E[Z^r]^{1/r} \leqq v$ $(1<r<\infty)$ <u>and let</u> r' <u>be the exponent conjugate to</u> r. <u>Then we have for</u> $\alpha \in]0,1[$

(72.7) $\quad P\{Z \geqq \alpha u\} \geqq (1-\alpha)^{r'} \dfrac{u^{r'}}{v^{r'}}$

PROOF. $\displaystyle \int_{\{Z \geqq \alpha u\}} Z = \int Z - \int_{\{Z < \alpha u\}} Z \geqq (1-\alpha)u$

$$\int_{\{Z \geqq \alpha u\}} Z \leqq E[Z^r]^{1/r} (P\{Z \geqq \alpha u\})^{1/r'}$$

by HÖLDER's inequality. The remainder is obvious.

We apply this result to $Z = e^{Y_\infty}$: to get the L^r estimate we just replace λ by λr . Dropping the conditional expectations for simplicity, we find

$$(72.8)\quad P\{Y_\infty \geqq \frac{\log \alpha + A\log(1+\varepsilon(-\lambda))}{\lambda}\} \geqq (1-\alpha)^{r'} \frac{e^{r'A\log(1+\varepsilon(-\lambda))}}{e^{r'\frac{\varepsilon(r\lambda)}{r}B}}$$

We **keep** α fixed, take $\lambda = 2\sqrt{c\log m} / p^m \sqrt{p^2-1}$, with $a<c<1/2$. Then $\lambda \to 0$ as $m \to \infty$, and therefore $\varepsilon(\lambda)$, $\varepsilon(-\lambda)$, $\log(1+\varepsilon(-\lambda))$ " behave" like $\lambda^2/2$ (let the reader scrape the rost from his elementary calculus, and write the details for himself). The left side of (72.8) then is smaller than

$$P\{Y_\infty \geqq p^m \sqrt{p^2-1} \sqrt{a\log m} \}$$

for m large enough. On the other hand, the right side " behaves" like

$$(1-\alpha)^{r'} \frac{e^{2r'c\log m}}{e^{2rr'c\log m}}$$

we replace rr' by $r+r'$, and we get a divergent series if $2rc<1$, which can be achieved since $c<1/2$. Here too, let the reader write the necessary details.

APPENDIX 1

DOOB'S PROOF OF THE MARTINGALE CONVERGENCE THEOREM

It may interest the reader to compare the proofs of the martingale convergence theorem given in chapter II to DOOB's original proof , which is far from obsolete (and might still be the best one for classes !). Let (X_n) be a martingale , and choose two numbers $a<b$. Assume first that $X \in L^1$ closes the martingale, and start with the maximal inequality (5.6) : if $C \in \underline{\underline{F}}_0$

(1) $$bP(C \cap \{\sup_{0 \leqq k \leqq n} X_k > b\}) \leqq \int_{C \cap \{\dots\}} X$$

Consider the following sets : for $m \leqq n$

$$B_{mn} = \{\sup_{m \leqq k \leqq n} X_k > b\} \quad , \quad A_{mn} = \{\inf_{m \leqq k \leqq n} X_k < a\}$$

Set also $D = \{\overline{\lim}_k X_k > b \text{ , } \underline{\lim}_k X_k < a\}$. Choose $k \leqq \ell \leqq m \leqq n$, and shift 0 to m in (1) : if $C \in \underline{\underline{F}}_m$

(2) $$bP(C \cap B_{mn}) \leqq \int_{C \cap B_{mn}} X$$

Take $C = A_{k\ell}$, then let successively n, m, ℓ, k tend to $+\infty$. There comes

(3) $$bP(D) \leqq \int_D X$$

Applying this result to $-X$ and $-a$ we get $aP(D) \geqq \int_D X$, which is absurd unless $P(D)=0$. Nothing could be simpler than this proof !

The proof for the L^1-bounded case is a little more complicated. Assume $P(D)>0$, and construct inductively integers $n_1 < n_2 \dots$ such that

$$P(D \cap A_{1n_1}) \geqq P(D)(1-4^{-1})$$
$$P(D \cap A_{1n_1} \cap B_{n_1 n_2}) \geqq P(D)(1-4^{-1}-4^{-2})$$
$$P(D \cap A_{1n_1} \cap B_{n_1 n_2} \cap A_{n_2 n_3}) \geqq P(D)(1-4^{-1}-4^{-2}-4^{-3}) \text{ ......}$$

Set $D_1 = A_{1n_1}$, $D_2 = A_{1n_1} \cap B_{n_1 n_2}$,... Then according to (2) or the similar inequality for inf instead of sup , $\int_{D_k} X_j$ is $\geqq bP(D_k)$ or $\leqq aP(D_k)$ for $j > n_k$, according as k is even or odd^k . Therefore $\int_{D_k \setminus D_{k+1}} X_j \geqq (b-a)P(D_{k+1}) > \frac{2}{3}(b-a)P(D)$ if $j > n_{k+1}$. This implies $E[|X_j|] \geqq \frac{2}{3} k(b-a)P(D)$, a contradiction to L^1-boundedness unless $P(D) =0$. The idea obviously is very close to that of the proof "with upcrossings", but one works on probabilities only, without looking at the sample paths.

APPENDIX 2

A PROOF OF THE BURKHOLDER INEQUALITIES (INCLUDING DAVIS's THEOREM)

We hope to include in this book some of the recent theory of H^p spaces, but it is probably too early now, since most of the material is still unpublished. On the other hand, while working on these H^p spaces, GETOOR and SHARPE discovered a new proof of the BURKHOLDER inequalities, which includes the more difficult case p=1 (DAVIS's theorem). Originally it was written only for sample continuous martingales, but the extension to right continuous ones (and hence to the discrete case) isn't difficult.

It is impossible to assert whether their proof is " definitive" - probably not ! Since this is just an appendix, and no proof of DAVIS's theorem has ever been published (as far as we know) in the right continuous case, we shall consider right continuous martingales Y relative to a family $(\underline{\underline{F}}_t)$ which satisfies the usual conditions. We shall assume that $Y_0=0$: if this isn't assumed, Y_0^2 must be added in the inequalities to the two increasing processes $[Y,Y]$ and $\langle Y,Y\rangle$. These processes, as well as stochastic integrals $V\cdot Y$ for predictable V, will be defined in the next volume. To apply the results and proofs to the discrete case , with a discrete family $\underline{\underline{F}}_n$, just set $\underline{\underline{F}}_t=\underline{\underline{F}}_n$ for $t\in[n,n+1[$, $Y_t=Y_n$, $V_t=V_n$ on the same interval. For t=n, $[Y,Y]_t$ and $\langle Y,Y\rangle_t$ are just the processes defined in n°54 . We set as usual $Y_t^* = \sup_{s\leq t}|Y_s|$, $Y^* = Y_\infty^*$.

We are going to prove the whole set of BURKHOLDER inequalities, but our advice is to use this proof only for $p\in]1,2[$ (or $]0,2[$). We feel that the inequalities for $p\geq 2$ aren't so deep as for $p<2$, and should be proved by a method like that of n°57. Some remarks on this subject are given at the end of appendix 2.

We are going to follow the original path of DAVIS, and start with the case of a martingale with " predictably bounded jumps" , which also gives very interesting side results concerning $\langle Y,Y\rangle$, and also the range $0<p<1$. We shall not try to cover the case of tame functions (n°s II.55 and 56) though it can probably be done .

1. Some auxiliary results about integration by parts will be necessary. Let A_t and B_t be two right continuous functions of bounded variation on $\mathbb{R}_+$. Then the integration by parts formula is

(1) $$d(A_tB_t) = A_{t-}dB_t+B_tdA_t \quad \text{(shortly : } d(AB)=A_-dB+BdA \text{)}$$

This formula has some amusing applications : for instance, assuming B is bounded from below (locally) and taking $A=1/B$ we get

$$d(\frac{1}{B}) = \frac{-dB}{B_-B}$$

For positive powers, we get

$$(3) \qquad dB^2 = (B_-+B)dB$$
$$dB^3 = (B_-^2+B_-B+B^2)dB$$
$$dB^4 = (B_-^3+B_-^2B+B_-B^2+B^3)dB \ldots$$

From which in turn some fractional powers can be deduced :

$$dB^{1/2} = \frac{dB}{B_-^{1/2}+B^{1/2}} \qquad dB^{1/3} = \frac{dB}{B_-^{2/3}+B_-^{1/3}B^{1/3}+B^{2/3}} \quad \ldots\ldots$$

We shall not need these equalities, but rather the following inequalities . Assume B is positive and increasing, and $0<r<1$. Then

$$(4) \qquad rB_-^{r-1}dB \geqq dB^r \geqq rB^{r-1}dB$$

inequalities being reversed if $r>1$, or if B is positive and decreasing. It is sufficient to prove (4) when r is rational. Set $r=k/q$, $k<q$, $A=B^{1/q}$. Then (4) can be written

$$\frac{k}{q} A_-^{k-q}dA^q \geqq dA^k \geqq \frac{k}{q}A^{k-q}dA^q \quad \text{or} \quad kA_-^k dA^q \geqq qA^q dA^k \geqq kA^k dA^q$$

which can be deduced from the identities (3) and the fact that $A_- \leqq A$. Integrating by parts the relation $B^r=B^{r-1}.B$ and using (4), we find that

$$(5) \qquad -B_-dB^{r-1} \leqq (\frac{1}{r}-1)dB^r \leqq -BdB^{r-1} \qquad (0<r<1) .$$

2. THEOREM. Let Y be a martingale (with $Y_0=0$) , D be an increasing process (we allow D_0 to be $\geqq 0$ instead of just $=0$). Assume that for all t

$$(6) \qquad |\Delta Y_t| = |Y_t - Y_{t-}| \leqq D_{t-}$$

Then we have for all $r\in]0,\infty[$

$$c_r E[[Y,Y]^r_\infty] \leqq E[(Y^*+D_\infty)^{2r}] \quad , \quad E[(Y^*)^{2r}] \leqq C'_r E[([Y,Y]_\infty + D^2_\infty)^r]$$

The same results are true for $\langle Y,Y\rangle$ instead of $[Y,Y]$.

PROOF. We are going to assume that all processes concerned are bounded, so as to avoid integrability difficulties. To go from this case to the general situation, consider the stopping time

$$T = \inf \{ t : |Y_t| \geqq N \text{ or } D_t \geqq N \text{ or } [Y,Y]_t \geqq N \}$$

Then Y, D, [Y,Y] are bounded on [0,N[; on the other hand the jump of Y at T is bounded by $D_{T-} \leqq N$, hence the processes Y and [Y,Y] are bounded on [0,N] . The process $(D_t \wedge N)$ dominates the jumps of the stopped martingale Y^T , $[Y^T,Y^T]$ and Y^T are bounded. We apply the theorem to them and let $N \to \infty$.

To have simple notations, we set $[Y,Y]_s = B_s$.

<u>First inequality</u>, $0<r<1$. We write

$$B^r_\infty = B^r_\infty (\varepsilon + D_\infty + Y^*)^{-2r(1-r)} . (\varepsilon + D_\infty + Y^*)^{2r(1-r)}$$

where ε is a constant >0 to spare us trouble with negative exponents. We apply HÖLDER's inequality with exponents 1/r and 1/1-r :

(7) $$E[B^r_\infty] \leqq E^r[B_\infty (\varepsilon + D_\infty + Y^*)^{2(r-1)}].E^{1-r}[(\varepsilon + D_\infty + Y^*)^{2r}]$$

The second term on the right side is ready to use. Since r<1 we have

$$B_\infty (\varepsilon + D_\infty + Y^*)^{2(r-1)} \leqq \int_0^\infty (\varepsilon + D_{s-} + Y^*_{s-})^{2(r-1)} dB_s$$

Consider the predictable process $V_s = (\varepsilon + D_{s-} + Y^*_{s-})^{r-1}$, and the martingale $X = V \cdot Y$: the last integral can be interpreted as $[X,X]_\infty$, and therefore the first term on the right side of (7) is smaller than

(8) $$E^r[[X,X]_\infty] = E^r[X^2_\infty]$$

We compute X_∞ with the help of the integration by parts formula for stochastic integrals (a triviality in the discrete case).

(9) $$X_\infty = \int_0^\infty (\varepsilon + D_{s-} + Y^*_{s-})^{r-1} dY_s = (\varepsilon + D_\infty + Y^*)^{r-1} Y_\infty - \int_0^\infty Y_s d(\varepsilon + D_s + Y^*_s)^{r-1}$$

$$|X_\infty| \leqq (\varepsilon + D_\infty + Y^*)^r - \int_0^\infty (\varepsilon + D_{s-} + Y^*_{s-}) d(\varepsilon + D_s + Y^*_s)^{r-1}$$

since $|Y_s| = |Y_{s-} + \Delta Y_s| \leqq \varepsilon + Y^*_{s-} + D_{s-}$. Applying (5) to $\varepsilon + Y^*_s + D_s$

(10) $$|X_\infty| \leqq \frac{1}{r} (\varepsilon + D_\infty + Y^*)^r$$

Coming back to (8) and (7), and letting $\varepsilon \to 0$, we have proved

(11) $$E[[Y,Y]_\infty] \leqq r^{-2r} E[(D_\infty + Y^*)^{2r}]$$

One may replace [Y,Y] by <Y,Y> all the way long.

<u>First inequality</u> , $r \geqq 1$. We write the inequality corresponding to (4) for the case $r \geqq 1$:

$$B^r_\infty \leqq r \int_0^\infty B^{r-1}_s dB_s \leqq r \int_0^\infty C^{r-1}_{s-} dB_s$$

where $C_s = B_s + D^2_s$, since the jumps of B are dominated by D^2_{s-} .

Consider the martingale $X = C_-^{(r-1)/2}.Y$. The last integral can be interpreted as $[X,X]_\infty$, and therefore

$$E[B^r_\infty] \leqq rE[[X,X]_\infty] = rE[X^2_\infty]$$

We compute X_∞ by integration by parts :

$$X_\infty = Y_\infty C_\infty^{(r-1)/2} - \int_0^\infty Y_s dC_s^{(r-1)/2} \leqq 2Y^* C_\infty^{(r-1)/2}$$

Therefore $E[B^r_\infty] \leqq 4rE[Y^{*2}C_\infty^{r-1}]$. We apply HÖLDER's inequality with exponents r and $r/r-1$:

$$\|B_\infty\|_r^r \leqq 4r\|Y^*\|_{2r}^2 \|C_\infty\|_r^{r-1}$$

If the martingale happens to be continuous, $C_\infty = B_\infty$, and the inequality follows at once. Otherwise, we apply MINKOWSKI's inequality

$$\|B_\infty\|_r^r \leqq 4r\|Y^*\|_{2r}^2(\|B_\infty\|_r + \|D_\infty\|_{2r}^2)^{r-1}$$

Assume $\|Y^*+D_\infty\|_{2r} \leqq 1$: since these r.v. are positive, their $\|\ \|_{2r}$ norms must both be $\leqq 1$, and we have

$$\frac{\|B_\infty\|_r^r}{(1+\|B_\infty\|_r)^{r-1}} \leqq 4r$$

which implies (since the function on the left side tends to $+\infty$ at infinity) that $\|B_\infty\|_r$ is bounded by some constant C.

We can again replace $[Y,Y]$ by $\langle Y,Y\rangle$ (some care is needed to check that the jump of $\langle Y,Y\rangle$ at s is dominated by D^2_{s-}).

<u>Second inequality</u> , $0<r<1$. Set as above $C_s=[Y,Y]_s+D_s^2$ ($\varepsilon>0$ may be added to avoid difficulties with negative exponents, but this isn't really necessary), and consider the martingale $X=C_-^{(r-1)/2}.Y$.
$[X,X]_\infty$ is $\int_0^\infty C_{s-}^{r-1}dB_s \leqq \int_0^\infty B_s^{r-1}dB_s \leqq \frac{1}{r}B^r_\infty$ (we recall that $r-1<0$).
According to DOOB's inequality II.8, we have

$$E[X^{*2}] \leqq 4E[X^2_\infty] = 4E[[X,X]_\infty] \leqq \frac{4}{r}E[B^r_\infty]$$

On the other hand, we have $Y=C_-^{(1-r)/2}.X$. Integrating by parts as above, we find that $|Y^*| \leqq 2C_\infty^{(1-r)/2}X^*$. Applying HÖLDER's inequality to the exponents $1/r$, $1/1-r$

$$E[(Y^*)^{2r}] \leqq 2^{2r}E[C_\infty^{r(1-r)}(X^*)^{2r}] \leqq 2^{2r}E^{1-r}[C^r_\infty]E^r[X^{*2}]$$
$$\leqq 2^{4r}r^{-r}E^{1-r}[C^r_\infty]E^r[B^r_\infty]$$

Replacing B^r_∞ by C^r_∞ , which is larger, gives us the result.
Here too, we may replace $[\,,\,]$ by $\langle\,,\,\rangle$.

<u>Second inequality</u>, $r \geqq 1$. Since this is the part of the proof for which the most advanced technique is used, let us point out at once that <u>it is not necessary</u> for the proof of the BURKHOLDER inequalities, as far as $[Y,Y]$ is concerned.

We "recall" the change of variables formula : if F is twice continuously differentiable on the line, and Y is as above

$$F(Y_\infty) = F(0) + \int_0^\infty F'(Y_{s-})dY_s + \frac{1}{2}\int_0^\infty F''(Y_{s-})\, d[Y^c,Y^c]_s$$

$$+ \sum_s (F(Y_s)-F(Y_{s-})-\Delta Y_s F'(Y_{s-}))$$

(we recall that $Y_0=0$, and Y is bounded). Here $[Y^c,Y^c]$ denotes the continuous part of $[Y,Y]$. We take $F(t)=|t|^{2r}$:

$$E[|Y_\infty|^{2r}] \leqq E[\ r(2r-1)\int_0^\infty |Y_s|^{2r-2} d[Y^c,Y^c]_s$$

$$+ \sum_s r(2r-1)\,\Delta Y_s^2 |Y_{s-}+\theta_s \Delta Y_s|^{2r-2}\ ]$$

We have applied Taylor's formula : θ_s belongs to $]0,1[$. Since $|\Delta Y_s| \leqq D_{s-}$, we may replace $|Y_{s-}+\theta_s\Delta Y_s|^{2r-2}$ by $(|Y_{s-}|+D_{s-})^{2r-2}$. Then, noticing that ΔY_s^2 is the jump of $[Y,Y]$ at s and putting together the continuous and discontinuous parts of $[Y,Y]$, we find that

$$E[|Y_\infty|^{2r}] \leqq r(2r-1)E[\int_0^\infty (|Y_{s-}|+D_{s-})^{2r-2} d[Y,Y]_s]$$

$$\leqq r(2r-1)E[\ (Y^*+D_\infty)^{2r-2}[Y,Y]_\infty\ ]$$

We apply HÖLDER's inequality with exponents r and $\frac{r}{r-1}$ to the right side, DOOB's inequality to the left side :

$$E[(Y^*)^{2r}] \leqq \left(\frac{2r}{2r-1}\right)^{2r} E[Y_\infty^{2r}] \leqq \left(\frac{2r}{2r-1}\right)^{2r} r(2r-1)E^{1/r}[[Y,Y]_\infty^r]\ E^{1-1/r}[(Y^*+D_\infty)^{2r}]$$

Assume $\|[Y,Y]_\infty + D_\infty^2\|_r \leqq 1$: it isn't difficult to see, as in the second part of the proof, that $E[(Y^*)^{2r}]$ is bounded by some constant C. The theorem is proved.

The same proof works with $\langle Y,Y\rangle$, just because $E[\int_0^\infty V_s d[Y,Y]_s] = E[\int_0^\infty V_s d\langle Y,Y\rangle_s]$ for any <u>predictable</u> process $V \geqq 0$.

COMMENT. If we want only to prove the BURKHOLDER inequalities, and DAVIS's theorem, then, as we shall see later on, we need only consider the case $r \in]0,1[$. However, this theorem shouldn't just be considered as a technical lemma to prove the BURKHOLDER inequalities.

1) In the range $0<2r<1$, it gives results concerning $[Y,Y]$ which aren't contained in the (usual) BURKHOLDER inequalities.

2) In the whole range $0<2r<\infty$, it gives information concerning the process $<Y,Y>$ which seems quite valuable. Compare for instance with n°s 56,58 , which concern the case of a process $D_s=1$: we are not considering tame functions, just powers, but the estimates are much better, and extend to some non convex functions (powers with exponents <1).

DAVIS's DECOMPOSITION AND BURKHOLDER's INEQUALITIES

3 . Let Y be a right continuous martingale ; we assume that $Y_0=0$ as above, though such a condition is quite unimportant. We set

(3.1) $$S_t = \sup_{s\leqq t} |\Delta Y_s|$$

If t is such that $|\Delta Y_t| \geqq 2S_{t-}$, we have

(3.2) $$|\Delta Y_t| + 2S_{t-} \leqq 2|\Delta Y_t| = 2S_t$$

Hence $|\Delta Y_t| \leqq 2(S_t-S_{t-})$, and the process

(3.3) $$K_t = \sum_{s\leqq t\ ,\ |\Delta Y_s|\geqq 2S_{t-}} \Delta Y_s$$

has a total variation at most equal to $2S_\infty$. If either Y^* or $[Y,Y]_\infty^{1/2}$ belongs to L^1, S_∞ is integrable, and we shall see in volume 2 that the compensated process $\widetilde{K}_t$ can be suitably defined, as the only predictable process of integrable variation (i.e. difference of two predictable, integrable increasing processes) such that $K^c=K-\widetilde{K}$ is a martingale. Of course this is trivial in the discrete case (I.18).

We decompose K into a difference of two increasing processes, each of total variation $\leqq 2S_\infty$, and apply to each of them the theorem of BURKHOLDER, DAVIS and GUNDY (II.53) , with $F(t)=t^p$, $1\leqq p < \infty$, or rather the extension of this theorem to the continuous case. Putting things together again, we get

(3.4) $$\|\text{total var. of } K^c\|_p \leqq C_p\|S_\infty\|_p \quad , \quad \|(K^c)^*\|_p \leqq C_p\|S_\infty\|_p$$

Here and below, C_p denotes some number which depends only on p, and can change from place to place.

Set $L=Y-K$, $\widetilde{L}=-\widetilde{K}$, $L^c=L-\widetilde{L}$, a martingale : these notations aren't quite correct, since L isn't a process of integrable variation, and hence hasn't any " compensated process". Let us prove that $|\Delta L^c_t| \leqq 4S_{t-}$ for all t : L^c has " predictably bounded jumps".

It is sufficient to show that $|\Delta L^c_T| \leqq 4S_{T-}$ for any stopping time T which is either totally inaccessible, or predictable. Since $|\Delta \underset{\sim}{L}_T| \leqq 2S_{T-}$ by definition of L , we just have to see whether $|\Delta \tilde{L}_T| = |\Delta \tilde{K}_T| \leqq 2S_{T-}$. This is certainly true if T is totally inaccessible, since $\tilde{K}$, being predictable, has no totally inaccessible jumps. If T is predictable, we have $|\Delta \tilde{L}_T| = |E[\Delta L_T | \underline{\underline{F}}_{T-}]| \leqq E[2S_{T-} | \underline{\underline{F}}_{T-}] = 2S_{T-}$, and the property is proved.

The decomposition $Y = K^c + L^c$ is called the DAVIS decomposition of Y. We are going to use it to prove the BURKHOLDER inequalities and the theorem of DAVIS.

4. PROOF OF THE INEQUALITIES

Inequality $\| [Y,Y]^{1/2}_\infty \|_p \leqq C_p \|Y^*\|_p$. We first write $[Y,Y]^{1/2}_\infty \leqq [K^c,K^c]^{1/2}_\infty + [L^c,L^c]^{1/2}_\infty$, so it suffices to consider separately the two norms $\| \; \|_p$ The first one is the norm of $(\sum \Delta K^{c\,2}_s)^{1/2} \leqq \sum |\Delta K^c_s|$, and is dealt with in formula (3.4), since $S_\infty \leqq 2Y^*$. To the second one we may apply the first inequality in theorem 2 :

$$\| [L^c,L^c]^{1/2}_\infty \|_p \leqq C_p \| L^{c*} + 4S_\infty \|_p$$

We dominate S_∞ by $2Y^*$, L^{c*} by $Y^* + K^{c*}$, and estimate $\|K^{c*}\|_p$ by (3.4). The inequality is proved.

Inequality $\|Y^*\|_p \leqq C_p \| [Y,Y]^{1/2}_\infty \|_p$. As above it suffices to dominate $\|K^{c*}\|_p$ and $\|L^{c*}\|_p$. For the first one we use (3.4), and remark that $S_\infty \leqq [Y,Y]^{1/2}_\infty$. For the second one we use the second inequality in theorem 2 :

$$\|L^{c*}\|_p \leqq C_p \| [L^c,L^c]^{1/2}_\infty + 4S_\infty \|_p$$

We dominate S_∞ by $[Y,Y]^{1/2}_\infty$, $[L^c,L^c]^{1/2}_\infty$ by $[Y,Y]^{1/2}_\infty + [K^c,K^c]^{1/2}_\infty$, and this last term by the total variation of K^c, to which we apply (3.4). The inequality follows.

5. COMMENT. The reader may have thought that the proof of theorem 2, consisting of four different parts, one of which uses the change of variables formula, isn't very pleasant... it may be reduced by more than one half in the following way.

One separates the BURKHOLDER inequalities in two sets : the range $1 \leqq p < 2$ ("deep") , and the range $2 \leqq p < \infty$ ("easy"). For the deep part (which includes DAVIS's theorem), nothing is changed. For the easy part, we have an easy proof in the direction

$$(5.1) \qquad \|[Y,Y]_\infty^{1/2}\|_p \leqq c_p \| Y^* \|_p$$

suggested in n°II.58, and we now have this inequality in the whole range $1 \leqq p < \infty$. But GARSIA[1] pointed out that this direction implies the reverse one, as follows : let X be any bounded right continuous martingale. We have (q denoting the exponent conjugate to p)

$$|E[X_\infty Y_\infty]| = |E[[X,Y]_\infty]| \leqq E[[X,X]_\infty^{1/2}[Y,Y]_\infty^{1/2}] \leqq$$
$$\leqq E^{1/p}[[Y,Y]_\infty^{p/2}]E^{1/q}[[X,X]_\infty^{q/2}]$$
$$\leqq c_p E^{1/p}[[Y,Y]_\infty^{p/2}]E^{1/q}[X^{*q}]$$
$$\leqq c_p E^{1/p}[[Y,Y]_\infty^{p/2}]E^{1/q}[|X_\infty|^q]$$

using (5.1) (for the exponent q) and DOOB's inequality II.8.Then we get $\|Y_\infty\|_p \leqq c_p \|[Y,Y]_\infty^{1/2}\|_p$, and a second application of DOOB's inequality gives the same result for Y^*. Of course, this reasoning doesn't apply to the theorem of DAVIS (p=1).

This approach to the inequalities doesn't use the MARCINKIEWICZ interpolation theorem, nor the RADEMACHER functions. On the other hand, if the method of n°s 48 and 60 is used, one sees that there is really no difference in depth between the ranges]1,2] and [2,∞[: they are dual ! It is only the theorem of DAVIS which plays a special role.

The conclusion seems to be that the subject is now beginning to be understood, but that much work remains to be done.

---

[1] Communicated by R.GETOOR. GARSIA's work is as yet unpublished.

APPENDIX 3

In his small book [32] on the LITTLEWOOD-PALEY theory, STEIN uses some extensions and consequences of the classical BURKHOLDER L^p inequalities. We are going to mention these results here (without proof) .

A RESULT RELATED TO THEOREM II.53

Theorem 53 asserts the following : if $b_0, b_1, \dots, b_n \dots$ are positive integrable random variables, and $a_i = E[b_i | \underline{\underline{F}}_i]$, then

(3.1) $$\| \sum a_i \|_p \leqq K_p \| \sum b_i \|_p \qquad (1 \leqq p < \infty)$$

Set now $b_i = f_i^2$, $f_i \in L^2$. It follows immediately that

(3.2) $$\| (\sum E^2 [f_i | \underline{\underline{F}}_i])^{1/2} \|_{2p} \leqq \| (\sum E[f_i^2 | \underline{\underline{F}}_i])^{1/2} \|_{2p}$$
$$K_p \| (\sum f_i^2)^{1/2} \|_{2p}$$

Here the exponent ranges over $[2, \infty[$. What can be said for the interval $]1,2[$? STEIN's result is the following

THEOREM. <u>For</u> $p \in]1,2]$ <u>the inequality</u>

(3.3) $$\| (\sum E^2 [f_i | \underline{\underline{F}}_i])^{1/2} \|_p \leqq c_p \| (\sum f_i^2)^{1/2} \|_p$$

<u>still is valid, and can even be replaced by</u>

(3.4) $$\| (\sum f_i^{*2})^{1/2} \|_p \leqq c_p \; \| (\sum f_i^2)^{1/2} \|_p$$

<u>where</u> $f_i^* = \sup_k \; | E[f_i | \underline{\underline{F}}_k] |$.

STEIN's proof ([32], p.103-105) isn't difficult, but uses an interpolation result of the RIESZ type , a method we haven't considered in the preceding chapters. I don't know whether the relation between STEIN's result and theorem 53 (and the similar theorems) has become clearer now.

MATRIX OPERATORS

Consider a martingale $X_0, \dots, X_n$ with a finite index set, and the corresponding difference sequence $x_0, \dots, x_n$. On the other hand, consider a (n,n) matrix (a_{ij}) - the coefficients in the general theory can be assumed to be "predictable", but we shall keep them fixed here and set

$$M_n = \left(\sum_{i=0}^{n} \left(\sum_{j=0}^{n} a_{ij} x_j \right)^2 \right)^{1/2}$$

Set also

$$(3.5) \qquad \begin{aligned} d_n^2 &= \inf_j \sum_{i=0}^{n} a_{ij}^2 \\ D_n^2 &= \sup_j \sum_{i=0}^{n} a_{ij}^2 \end{aligned}$$

Then we have the following theorem, due to BURKHOLDER-GUNDY (see [31], p.288-295 : they prove stronger results).

THEOREM. <u>For</u> $1<p<\infty$ <u>we have</u>

$$(3.6) \qquad d_n k_p \| Q_n^X \|_p \leqq \| M_n \|_p \leqq D_n K_p \| Q_n^X \|_p$$

<u>where</u> Q_n^X <u>is the quadratic variation</u>, <u>and</u> k_p, K_p <u>depend only on</u> p, <u>not on</u> n.

If $(a_{ij})$is the identity matrix, we get for M_n Q_n^X itself. The proof in [31] rests on the BURKHOLDER inequalities, and the lemma of KHINCHIN (II.59).

The main application of this theorem (which plays an essential role in STEIN's book) is to CESARO means of martingales or reversed martingales : set $C_0=\hat{C}_0=0$, and

$$\begin{aligned} &C_1 = X_0 &&, \quad \hat{C}_1 = X_n \\ &C_2 = (X_0+X_1)/2 &&, \quad \hat{C}_2 = (X_n+X_{n-1})/2 \\ &\cdots\cdots\cdots\cdots &&\quad \cdots\cdots\cdots \\ &C_{n+1} = (X_0+\ldots+X_n)/n+1 &&, \quad \hat{C}_{n+1} = (X_n+\ldots+X_0)/n+1 \end{aligned}$$

and set

$$M_n = \left(\sum_{k=1}^{n+1} k (C_k - C_{k-1})^2 \right)^{1/2}$$

$\hat{M}_n$ being defined in the same way, these two operators are matrix operators, and the corresponding coefficients D_n from (3.5) are uniformly bounded. Let us write the coefficients a_{ij} for M_n (the boundedness of D_n^2 is easy to get if the discrete sums are approximated by definite integrals)

$$\begin{aligned} &a_{0j} = \delta_{0j} \\ &a_{ij} = 0 \text{ if } j>i \ , \quad a_{ij} = \sqrt{i+1}\, \frac{j+1}{i(i+1)} \quad \text{if } j \leqq i \end{aligned}$$

NOTATIONS AND DEFINITIONS

PROVISORY BIBLIOGRAPHY

The items here appear in the order they are quoted in the book. Alphabetical order thus isn't respected !

[1]. MEYER (P.A.). Probabilités et potentiel. Blaisdell, Boston ; Hermann, Paris, 1966.

[2]. SAKS (S.). Theory of the integral. 2nd ed, translated by L.C. Young. Warszawa, Monografie Matematyczne n°7, 1937.

[3]. BURKHOLDER (D.L.). Martingale transforms. Ann.Math. Stat. 37, 1966, 1494-1504 .

[4]. DOOB (J.L.). Stochastic processes, Wiley 1953 .

[5]. DELLACHERIE (C.). Book on the general theory of processes, to appear in 1972 in the Ergebnisse (Springer) collection.

[6]. CORNEA (A.) and LICEA (G.). Une démonstration unifiée des théorèmes de sections. Z. für W-theorie, 10, 1968 ,p. 198-202

[7]. CHUNG (K.L.) and DOOB (J.L.). Fields, optionality and measurability. Amer. J. Math., 87, 1965, 397-424.

[8]. STEIN (E.M.). Singular integrals and differentiability properties of functions. Princeton University press, 1970.

[9]. BLACKWELL (D.) and DUBINS (L.E.). A converse to the dominated convergence theorem. Ill. J. Math.,7, 1963, p.508-514.

[10]. GUNDY (R.F.). On the class LlogL, martingales and singular integrals. Ann.Math.Stat., 39, 1968, p.134-138.

[11]. DOOB (J.L.). Regularity properties of certain families of chance variables. Trans. Amer.M. Soc.,47, 1940, p.455-486.

[12]. CHATTERJI (S.D.). A note on the convergence of Banach space valued martingales. Math.Ann., 153, 1964, p.142-149.

[13]. CHATTERJI (S.D.). Comments on the martingale convergence theorem. Symposium on probability methods in analysis (Loutraki). Lecture Notes in Mathematics vol.31, Springer-Verlag 1967.

[14]. SCALORA (F.). Abstract martingale convergence theorems. Pacific J. Math., 11, 1961, 347-374.

[15]. IONESCU TULCEA (A.) and IONESCU TULCEA (C.). Abstract ergodic theorems. Trans.Amer.M.Soc. 107, 1963, p.313-337.

[16]. SNELL (J.L.). Applications of martingale system theorems. Trans. Amer.M.Soc. 73, 1952, 293-312.

[17]. DUBINS (L.E.). A note on upcrossings of semimartingales. Ann.M.Stat., 37, 1966, p. 728.

[18]. DUBINS (L.E.). Rises and upcrossings of nonnegative martingales Illinois J.Math. 6, 1962, p.226-241.

[19]. DAVIS (B.) .A comparison test for martingale inequalities. Ann. M.Stat. 40, 1969, p.505-508.

[20]. MERTENS (J.F.). Processus stochastiques généraux et surmartingales. Z. für W-th., 1971

[21]. MERTENS (J.F). Strongly supermedian functions. To appear.

[22]. ITO (K.)and WATANABE (S.). Transformation of Markov processes by multiplicative functionals. Ann.Inst.Fourier, 15, 1965, p.15-30.

[23]. RAO (M.). On decomposition theorems of Meyer. M.Scand.,24,1969, p.66-78 (see also the next paper, p.79-92).

[24]. BURKHOLDER (D.L), DAVIS (B.J.) and GUNDY (R.F.). Integral inequalities for convex functions of operators on martingales. Proceedings of the 6-th Berkeley Symposium.

[25]. DAVIS (B.J.). On the integrability of the martingale square function. Israel J. of M. , 8, 1970, 187-190.

[26]. DUBINS (L.E) and FREEDMAN (D.A.). A sharper form of the Borel-Cantelli lemma and the strong law. Ann. M. Stat. 36, 1965, 800-807.

[27]. MAISONNEUVE (B.). Quelques martingales remarquables associées à une martingale continue. Publ. ISUP, 1970.

[28]. STOUT(W.F.). A martingale analogue of KOLMOGOROV's law of the iterated logarithm. Z. W-theorie.15, 1970, p.279-290.

[29]. STOUT (W.F.). Maximal inequalities and the law of the iterated logarithm. To appear.

[30]. R. GETOOR and M.SHARPE. Conformal martingales. To appear in Invent. Math., 1972.

[31]. BURKHOLDER (D.L.) and GUNDY (R.F.). Extrapolation and interpolation of quasi-linear operators on martingales. Acta Math., 124, 1970, p.250-304.

[32]. STEIN (E.M.). Topics in harmonic analysis related to the LITTLEWOOD-PALEY theory. Annals of Math. Studies, n°63, Princeton 1970.

Lecture Notes in Mathematics

Comprehensive leaflet on request

Vol. 111: K. H. Mayer, Relationen zwischen charakteristischen Zahlen. III, 99 Seiten. 1969. DM 16,–

Vol. 112: Colloquium on Methods of Optimization. Edited by N. N. Moiseev. IV, 293 pages. 1970. DM 18,–

Vol. 113: R. Wille, Kongruenzklassengeometrien. III, 99 Seiten. 1970. DM 16,–

Vol. 114: H. Jacquet and R. P. Langlands, Automorphic Forms on GL (2). VII, 548 pages. 1970. DM 24,–

Vol. 115: K. H. Roggenkamp and V. Huber-Dyson, Lattices over Orders I. XIX, 290 pages. 1970. DM 18,–

Vol. 116: Séminaire Pierre Lelong (Analyse) Année 1969. IV, 195 pages. 1970. DM 16,–

Vol. 117: Y. Meyer, Nombres de Pisot, Nombres de Salem et Analyse Harmonique. 63 pages. 1970. DM 16,–

Vol. 118: Proceedings of the 15th Scandinavian Congress, Oslo 1968. Edited by K. E. Aubert and W. Ljunggren. IV, 162 pages. 1970. DM 16,–

Vol. 119: M. Raynaud, Faisceaux amples sur les schémas en groupes et les espaces homogènes. III, 219 pages. 1970. DM 16,–

Vol. 120: D. Siefkes, Büchi's Monadic Second Order Successor Arithmetic. XII, 130 Seiten. 1970. DM 16,–

Vol. 121: H. S. Bear, Lectures on Gleason Parts. III, 47 pages. 1970. DM 16,–

Vol. 122: H. Zieschang, E. Vogt und H.-D. Coldewey, Flächen und ebene diskontinuierliche Gruppen. VIII, 203 Seiten. 1970. DM 16,–

Vol. 123: A. V. Jategaonkar, Left Principal Ideal Rings. VI, 145 pages. 1970. DM 16,–

Vol. 124: Séminare de Probabilités IV. Edited by P. A. Meyer. IV, 282 pages. 1970. DM 20,–

Vol. 125: Symposium on Automatic Demonstration. V, 310 pages. 1970. DM 20,–

Vol. 126: P. Schapira, Théorie des Hyperfonctions. XI, 157 pages. 1970. DM 16,–

Vol. 127: I. Stewart, Lie Algebras. IV, 97 pages. 1970. DM 16,–

Vol. 128: M. Takesaki, Tomita's Theory of Modular Hilbert Algebras and its Applications. II, 123 pages. 1970. DM 16,–

Vol. 129: K. H. Hofmann, The Duality of Compact Semigroups and C*-Bigebras. XII, 142 pages. 1970. DM 16,–

Vol. 130: F. Lorenz, Quadratische Formen über Körpern. II, 77 Seiten. 1970. DM 16,–

Vol. 131: A Borel et al., Seminar on Algebraic Groups and Related Finite Groups. VII, 321 pages. 1970. DM 22,–

Vol. 132: Symposium on Optimization. III, 348 pages. 1970. DM 22,–

Vol. 133: F. Topsøe, Topology and Measure. XIV, 79 pages. 1970. DM 16,–

Vol. 134: L. Smith, Lectures on the Eilenberg-Moore Spectral Sequence. VII, 142 pages. 1970. DM 16,–

Vol. 135: W. Stoll, Value Distribution of Holomorphic Maps into Compact Complex Manifolds. II, 267 pages. 1970. DM 18,–

Vol. 136: M. Karoubi et al., Séminaire Heidelberg-Saarbrücken-Strasbourg sur la K-Théorie. IV, 264 pages. 1970. DM 18,–

Vol. 137: Reports of the Midwest Category Seminar IV. Edited by S. MacLane. III, 139 pages. 1970. DM 16,–

Vol. 138: D. Foata et M. Schützenberger, Théorie Géométrique des Polynômes Eulériens. V, 94 pages. 1970. DM 16,–

Vol. 139: A. Badrikian, Séminaire sur les Fonctions Aléatoires Linéaires et les Mesures Cylindriques. VII, 221 pages. 1970. DM 18,–

Vol. 140: Lectures in Modern Analysis and Applications II. Edited by C. T. Taam. VI, 119 pages. 1970. DM 16,–

Vol. 141: G. Jameson, Ordered Linear Spaces. XV, 194 pages. 1970. DM 16,–

Vol. 142: K. W. Roggenkamp, Lattices over Orders II. V, 388 pages. 1970. DM 22,–

Vol. 143: K. W. Gruenberg, Cohomological Topics in Group Theory. XIV, 275 pages. 1970. DM 20,–

Vol. 144: Seminar on Differential Equations and Dynamical Systems, II. Edited by J. A. Yorke. VIII, 268 pages. 1970. DM 20,–

Vol. 145: E. J. Dubuc, Kan Extensions in Enriched Category Theory. XVI, 173 pages. 1970. DM 16,–

Vol. 146: A. B. Altman and S. Kleiman, Introduction to Grothendieck Duality Theory. II, 192 pages. 1970. DM 18,–

Vol. 147: D. E. Dobbs, Cech Cohomological Dimensions for Commutative Rings. VI, 176 pages. 1970. DM 16,–

Vol. 148: R. Azencott, Espaces de Poisson des Groupes Localement Compacts. IX, 141 pages. 1970. DM 16,–

Vol. 149: R. G. Swan and E. G. Evans, K-Theory of Finite Groups and Orders. IV, 237 pages. 1970. DM 20,–

Vol. 150: Heyer, Dualität lokalkompakter Gruppen. XIII, 372 Seiten. 1970. DM 20,–

Vol. 151: M. Demazure et A. Grothendieck, Schémas en Groupes I. (SGA 3). XV, 562 pages. 1970. DM 24,–

Vol. 152: M. Demazure et A. Grothendieck, Schémas en Groupes II. (SGA 3). IX, 654 pages. 1970. DM 24,–

Vol. 153: M. Demazure et A. Grothendieck, Schémas en Groupes III. (SGA 3). VIII, 529 pages. 1970. DM 24,–

Vol. 154: A. Lascoux et M. Berger, Variétés Kähleriennes Compactes. VII, 83 pages. 1970. DM 16,–

Vol. 155: Several Complex Variables I, Maryland 1970. Edited by J. Horváth. IV, 214 pages. 1970. DM 18,–

Vol. 156: R. Hartshorne, Ample Subvarieties of Algebraic Varieties. XIV, 256 pages. 1970. DM 20,–

Vol. 157: T. tom Dieck, K. H. Kamps und D. Puppe, Homotopietheorie. VI, 265 Seiten. 1970. DM 20,–

Vol. 158: T. G. Ostrom, Finite Translation Planes. IV. 112 pages. 1970. DM 16,–

Vol. 159: R. Ansorge und R. Hass. Konvergenz von Differenzenverfahren für lineare und nichtlineare Anfangswertaufgaben. VIII, 145 Seiten. 1970. DM 16,–

Vol. 160: L. Sucheston, Constributions to Ergodic Theory and Probability. VII, 277 pages. 1970. DM 20,–

Vol. 161: J. Stasheff, H-Spaces from a Homotopy Point of View. VI, 95 pages. 1970. DM 16,–

Vol. 162: Harish-Chandra and van Dijk, Harmonic Analysis on Reductive p-adic Groups. IV, 125 pages. 1970. DM 16,–

Vol. 163: P. Deligne, Equations Différentielles à Points Singuliers Reguliers. III, 133 pages. 1970. DM 16,–

Vol. 164: J. P. Ferrier, Seminaire sur les Algebres Complètes. II, 69 pages. 1970. DM 16,–

Vol. 165: J. M. Cohen, Stable Homotopy. V, 194 pages. 1970. DM 16,–

Vol. 166: A. J. Silberger, PGL_2 over the p-adics: its Representations, Spherical Functions, and Fourier Analysis. VII, 202 pages. 1970. DM 18,–

Vol. 167: Lavrentiev, Romanov and Vasiliev, Multidimensional Inverse Problems for Differential Equations. V, 59 pages. 1970. DM 16,–

Vol. 168: F. P. Peterson, The Steenrod Algebra and its Applications: A conference to Celebrate N. E. Steenrod's Sixtieth Birthday. VII, 317 pages. 1970. DM 22,–

Vol. 169: M. Raynaud, Anneaux Locaux Henséliens. V, 129 pages. 1970. DM 16,–

Vol. 170: Lectures in Modern Analysis and Applications III. Edited by C. T. Taam. VI, 213 pages. 1970. DM 18,–

Vol. 171: Set-Valued Mappings, Selections and Topological Properties of 2^X. Edited by W. M. Fleischman. X, 110 pages. 1970. DM 16,–

Vol. 172: Y.-T. Siu and G. Trautmann, Gap-Sheaves and Extension of Coherent Analytic Subsheaves. V, 172 pages. 1971. DM 16,–

Vol. 173: J. N. Mordeson and B. Vinograde, Structure of Arbitrary Purely Inseparable Extension Fields. IV, 138 pages. 1970. DM 16,–

Vol. 174: B. Iversen, Linear Determinants with Applications to the Picard Scheme of a Family of Algebraic Curves. VI, 69 pages. 1970. DM 16,–

Vol. 175: M. Brelot, On Topologies and Boundaries in Potential Theory. VI, 176 pages. 1971. DM 18,–

Vol. 176: H. Popp, Fundamentalgruppen algebraischer Mannigfaltigkeiten. IV, 154 Seiten. 1970. DM 16,–

Vol. 177: J. Lambek, Torsion Theories, Additive Semantics and Rings of Quotients. VI, 94 pages. 1971. DM 16,–

Please turn over

Vol. 178: Th. Bröcker und T. tom Dieck, Kobordismentheorie. XVI, 191 Seiten. 1970. DM 18,–

Vol. 179: Seminaire Bourbaki – vol. 1968/69. Exposés 347-363. IV, 295 pages. 1971. DM 22,–

Vol. 180: Séminaire Bourbaki – vol. 1969/70. Exposés 364-381. IV, 310 pages. 1971. DM 22,–

Vol. 181: F. DeMeyer and E. Ingraham, Separable Algebras over Commutative Rings. V, 157 pages. 1971. DM 16.–

Vol. 182: L. D. Baumert. Cyclic Difference Sets. VI, 166 pages. 1971. DM 16,–

Vol. 183: Analytic Theory of Differential Equations. Edited by P. F. Hsieh and A. W. J. Stoddart. VI, 225 pages. 1971. DM 20,–

Vol. 184: Symposium on Several Complex Variables, Park City, Utah, 1970. Edited by R. M. Brooks. V, 234 pages. 1971. DM 20,–

Vol. 185: Several Complex Variables II, Maryland 1970. Edited by J. Horváth. III, 287 pages. 1971. DM 24,–

Vol. 186: Recent Trends in Graph Theory. Edited by M. Capobianco/J. B. Frechen/M. Krolik. VI, 219 pages. 1971. DM 18.–

Vol. 187: H. S. Shapiro, Topics in Approximation Theory. VIII, 275 pages. 1971. DM 22,–

Vol. 188: Symposium on Semantics of Algorithmic Languages. Edited by E. Engeler. VI, 372 pages. 1971. DM 26,–

Vol. 189: A. Weil, Dirichlet Series and Automorphic Forms. V. 164 pages. 1971. DM 16,–

Vol. 190: Martingales. A Report on a Meeting at Oberwolfach, May 17-23, 1970. Edited by H. Dinges. V, 75 pages. 1971. DM 16,–

Vol. 191: Séminaire de Probabilités V. Edited by P. A. Meyer. IV, 372 pages. 1971. DM 26,–

Vol. 192: Proceedings of Liverpool Singularities – Symposium I. Edited by C. T. C. Wall. V, 319 pages. 1971. DM 24,–

Vol. 193: Symposium on the Theory of Numerical Analysis. Edited by J. Ll. Morris. VI, 152 pages. 1971. DM 16,–

Vol. 194: M. Berger, P. Gauduchon et E. Mazet. Le Spectre d'une Variété Riemannienne. VII, 251 pages. 1971. DM 22,–

Vol. 195: Reports of the Midwest Category Seminar V. Edited by J.W. Gray and S. Mac Lane. III, 255 pages. 1971. DM 22,–

Vol. 196: H-spaces – Neuchâtel (Suisse)- Août 1970. Edited by F. Sigrist, V, 156 pages. 1971. DM 16,–

Vol. 197: Manifolds – Amsterdam 1970. Edited by N. H. Kuiper. V, 231 pages. 1971. DM 20,–

Vol. 198: M. Hervé, Analytic and Plurisubharmonic Functions in Finite and Infinite Dimensional Spaces. VI, 90 pages. 1971. DM 16.–

Vol. 199: Ch. J. Mozzochi, On the Pointwise Convergence of Fourier Series. VII, 87 pages. 1971. DM 16,–

Vol. 200: U. Neri, Singular Integrals. VII, 272 pages. 1971. DM 22,–

Vol. 201: J. H. van Lint, Coding Theory. VII, 136 pages. 1971. DM 16,–

Vol. 202: J. Benedetto, Harmonic Analysis on Totally Disconnected Sets. VIII, 261 pages. 1971. DM 22,–

Vol. 203: D. Knutson, Algebraic Spaces. VI, 261 pages. 1971. DM 22,–

Vol. 204: A. Zygmund, Intégrales Singulières. IV, 53 pages. 1971. DM 16,–

Vol. 205: Séminaire Pierre Lelong (Analyse) Année 1970. VI, 243 pages. 1971. DM 20,–

Vol. 206: Symposium on Differential Equations and Dynamical Systems. Edited by D. Chillingworth. XI, 173 pages. 1971. DM 16,–

Vol. 207: L. Bernstein, The Jacobi-Perron Algorithm – Its Theory and Application. IV, 161 pages. 1971. DM 16,–

Vol. 208: A. Grothendieck and J. P. Murre, The Tame Fundamental Group of a Formal Neighbourhood of a Divisor with Normal Crossings on a Scheme. VIII, 133 pages. 1971. DM 16,–

Vol. 209: Proceedings of Liverpool Singularities Symposium II. Edited by C. T. C. Wall. V, 280 pages. 1971. DM 22,–

Vol. 210: M. Eichler, Projective Varieties and Modular Forms. III, 118 pages. 1971. DM 16,–

Vol. 211: Théorie des Matroïdes. Edité par C. P. Bruter. III, 108 pages. 1971. DM 16,–

Vol. 212: B. Scarpellini, Proof Theory and Intuitionistic Systems. VII, 291 pages. 1971. DM 24,–

Vol. 213: H. Hogbe-Nlend, Théorie des Bornologies et Applications. V, 168 pages. 1971. DM 18,–

Vol. 214: M. Smorodinsky, Ergodic Theory, Entropy. V, 64 pages. 1971. DM 16,–

Vol. 215: P. Antonelli, D. Burghelea and P. J. Kahn, The Concordance-Homotopy Groups of Geometric Automorphism Groups. X, 140 pages. 1971. DM 16,–

Vol. 216: H. Maaß, Siegel's Modular Forms and Dirichlet Series. VII, 328 pages. 1971. DM 20,–

Vol. 217: T. J. Jech, Lectures in Set Theory with Particular Emphasis on the Method of Forcing. V, 137 pages. 1971. DM 16,–

Vol. 218: C. P. Schnorr, Zufälligkeit und Wahrscheinlichkeit. IV, 212 Seiten 1971. DM 20,–

Vol. 219: N. L. Alling and N. Greenleaf, Foundations of the Theory of Klein Surfaces. IX, 117 pages. 1971. DM 16,–

Vol. 220: W. A. Coppel, Disconjugacy. V, 148 pages. 1971. DM 16,–

Vol. 221: P. Gabriel und F. Ulmer, Lokal präsentierbare Kategorien. V, 200 Seiten. 1971. DM 18,–

Vol. 222: C. Meghea, Compactification des Espaces Harmoniques. III, 108 pages. 1971. DM 16,–

Vol. 223: U. Felgner, Models of ZF-Set Theory. VI, 173 pages. 1971. DM 16,–

Vol. 224: Revètements Etales et Groupe Fondamental. (SGA 1). Dirigé par A. Grothendieck XXII, 447 pages. 1971. DM 30,–

Vol. 225: Théorie des Intersections et Théorème de Riemann-Roch. (SGA 6). Dirigé par P. Berthelot, A. Grothendieck et L. Illusie. XII, 700 pages. 1971. DM 40,–

Vol. 226: Seminar on Potential Theory, II. Edited by H. Bauer. IV, 170 pages. 1971. DM 18,–

Vol. 227: H. L. Montgomery, Topics in Multiplicative Number Theory. IX, 178 pages. 1971. DM 18,–

Vol. 228: Conference on Applications of Numerical Analysis. Edited by J. Ll. Morris. X, 358 pages. 1971. DM 26,–

Vol. 229: J. Väisälä, Lectures on n-Dimensional Quasiconformal Mappings. XIV, 144 pages. 1971. DM 16,–

Vol. 230: L. Waelbroeck, Topological Vector Spaces and Algebras. VII, 158 pages. 1971. DM 16,–

Vol. 231: H. Reiter, L^1-Algebras and Segal Algebras. XI, 113 pages. 1971. DM 16,–

Vol. 232: T. H. Ganelius, Tauberian Remainder Theorems. VI, 75 pages. 1971. DM 16,–

Vol. 233: C. P. Tsokos and W. J. Padgett. Random Integral Equations with Applications to Stochastic Systems. VII, 174 pages. 1971. DM 18,–

Vol. 234: A. Andreotti and W. Stoll. Analytic and Algebraic Dependence of Meromorphic Functions. III, 390 pages. 1971. DM 26,–

Vol. 235: Global Differentiable Dynamics. Edited by O. Hájek, A. J. Lohwater, and R. McCann. X, 140 pages. 1971. DM 16,–

Vol. 236: M. Barr, P. A. Grillet, and D. H. van Osdol. Exact Categories and Categories of Sheaves. VII, 239 pages. 1971. DM 20,–

Vol. 237: B. Stenström. Rings and Modules of Quotients. VII, 136 pages. 1971. DM 16,–

Vol. 238: Der kanonische Modul eines Cohen-Macaulay-Rings. Herausgegeben von Jürgen Herzog und Ernst Kunz. VI, 103 Seiten. 1971. DM 16,–

Vol. 239: L. Illusie, Complexe Cotangent et Déformations I. XV, 355 pages. 1971. DM 26,–

Vol. 240: A. Kerber, Representations of Permutation Groups I. VII, 192 pages. 1971. DM 18,–

Vol. 241: S. Kaneyuki, Homogeneous Bounded Domains and Siegel Domains. V, 89 pages. 1971. DM 16,–

Vol. 242: R. R. Coifman et G. Weiss, Analyse Harmonique Non-Commutative sur Certains Espaces. V, 160 pages. 1971. DM 16,–

Vol. 243: Japan-United States Seminar on Ordinary Differential and Functional Equations. Edited by M. Urabe. VIII, 332 pages. 1971. DM 26,–

Vol. 244: Séminaire Bourbaki – vol. 1970/71. Exposés 382–399. IV, 356 pages. 1971. DM 26,–

Vol. 245: D. E. Cohen, Groups of Cohomological Dimension One. V, 99 pages. 1972. DM 16,–

MIX
Papier aus verantwortungsvollen Quellen
Paper from responsible sources
FSC® C105338

If you have any concerns about our products,
you can contact us on
ProductSafety@springernature.com

In case Publisher is established outside the EU,
the EU authorized representative is:
Springer Nature Customer Service Center GmbH
Europaplatz 3, 69115 Heidelberg, Germany

Printed by Libri Plureos GmbH
in Hamburg, Germany